THE
UNNATURAL
WORLD

The Race to Remake Civilization in Earth's Newest Age

DAVID BIELLO

SCRIBNER

New York   London   Toronto   Sydney   New Delhi

Scribner
An Imprint of Simon & Schuster, Inc.
1230 Avenue of the Americas
New York, NY 10020

First Scribner hardcover edition November 2016

SCRIBNER and design are registered trademarks of The Gale Group, Inc., used under license by Simon & Schuster, Inc., the publisher of this work.

For information about special discounts for bulk purchases, please contact Simon & Schuster Special Sales at 1-866-506-1949 or business@simonandschuster.com.

The Simon & Schuster Speakers Bureau can bring authors to your live event. For more information or to book an event, contact the Simon & Schuster Speakers Bureau at 1-866-248-3049 or visit our website at www.simonspeakers.com.

Interior design by Jill Putorti

Manufactured in the United States of America

10   9   8   7   6   5   4   3   2   1

Library of Congress Cataloging-in-Publication Data

Names: Biello, David, author.
Title: The unnatural world : the race to remake civilization in Earth's
    newest age / David Biello.
Description: New York : Scribner, an Imprint of Simon & Schuster, Inc. [2016]
Identifiers: LCCN 2016018491 (print) | LCCN 2016031269 (ebook) | ISBN
    9781476743905 | ISBN 9781476743929 (ebook) | ISBN 9781476743929
Subjects: LCSH: Human ecology. | Nature and civilization. | Global
    environmental change. | Nature—Effect of human beings on. | Civilization,
    Modern—21st century.
Classification: LCC GF75 .B54 2016 (print) | LCC GF75 (ebook) | DDC
    304.2 —dc23
LC record available at https://lccn.loc.gov/2016018491

ISBN 978-1-4767-4390-5
ISBN 978-1-4767-4392-9 (ebook)

For my own Anthropocene working group and inspirations:
Beatrice, Dan, Desmond, Elizabeth, and Shii Ann

Contents

The Overview   1

PART I: ALTER EARTH   9

Chapter 1: Iron Rules   11

Chapter 2: Written in Stone   39

PART II: TERRA INCOGNITA   67

Chapter 3: Ground Work   69

Chapter 4: Big Death   97

Chapter 5: The People's Epoch   131

PART III: A BETTER ANTHROPOCENE   163

Chapter 6: City Folks   165

Chapter 7: The Long Thaw   201

Chapter 8: The Final Frontier   232

Acknowledgments   267

Selected Notes and Further Reading   271

Index   277

The Overview

The view from space changes people.

Just before Christmas 1968, William "Bill" Anders and two other men spun around the moon in the prosaically named Lunar Module, travelers on a course as far from Earth as any human has ever been.

Apollo 8 launched from Cape Canaveral on the morning of December 21. As Anders, Commander Frank Borman, and Module Pilot James Lovell blasted through and beyond the atmosphere, Earth dwindled from planet-size to basketball-size. The astronauts busied themselves with the flight checklist. By the time eleven hours had passed, the planet appeared no bigger than a baseball, and within three days the astronauts had floated all the way to the moon.

At precisely the moment NASA planners had predicted, this lonely manned module in the vastness of space lost radio contact. The astronauts passed behind the moon, and the entire Earth disappeared. The men circled our planet's steadfast companion three times, diligently recording its pockmarked, colorless surface, littered with the blast craters and detritus of bombarding asteroids. From roughly 100 kilometers up, the moon didn't look like much. It reminded Anders of "a deserted beach that had been churned by footprints during a volleyball game" or, as he put it later in the flight, "beach sand darkened by the cold embers of bonfires," which earned him hate mail from poets.

Then, on their fourth orbit, a revelation. Before the Apollo 8 astronauts' eyes, home slid out from behind the horizon—small enough to be blotted out by a thumb.

"Oh my God! Look at that picture over there!" Anders cried out. What he caught glimpse of, as Borman later described, was simply "the most beautiful, heart-catching sight" of their lives, "one that sent a torrent of nostalgia, of sheer homesickness, surging through" them.

"This," Anders thought, "must be what God sees."

Anders raised his camera—over the protest of Borman, who fretted about taking an unscheduled photo—and snapped the picture that would make this mission famous and change the human perspective on the world forever.

As the British physicist Fred Hoyle predicted back in 1950, a picture that revealed the whole Earth shining in the empty darkness of space would unleash "a new idea as powerful as any in history." That idea is now known as the Anthropocene, the notion that humanity has become a world-changing force of nature. That Earth is in the grip of human nature.

Most of us have never been to space and, despite the best efforts of industrialists and dreamers like Elon Musk, most of us never will. The closest I've come is on flights across continents, looking over green irrigation circles blooming in the desert or the snowy wastes that edge the Arctic. But anyone with an Internet connection can now enjoy NASA pictures and Google Earth views. Only twenty-four men have left Earth's orbit, but satellites have made millions of us voyeurs of the planet.

The first such group was actually a television audience, those in the United States and elsewhere who tuned in on December 22, 1968, when the Apollo 8 astronauts turned a camera through the window for a live view. As Earth appeared on the television screen, Anders told this audience, "You're looking at yourselves as seen from 180,000 miles out in space."

Anders was not an especially religious man, but during a subsequent Christmas Eve broadcast from space, he left viewers with this message from the Bible, the opening lines of Genesis: "In the beginning God created the heavens and the earth. And the earth was without form, and void; and darkness was upon the face of the deep. And the spirit of God moved upon the face of the waters. And God said, 'Let there be light,' and there was light. And God saw the light, that it was good and God divided the light from the darkness."

Like a blue-and-white jewel set in the firmament, Earth gleams in the inky darkness of space—our only home. Just as looking in the mirror can reveal our own flaws and strengths, looking back at the planet provided a fresh perspective. We humans are fragile animals, like most life on Earth larger than a single cell. We are trapped, utterly reliant on our spaceship-planet; without it death creeps in from the cold, indifferent void. Even so, it may not be space that kills us in the end, though there's always the chance of another obliterating asteroid like the one that ended the reign of the dinosaurs. Instead, like the Soviet dog Laika, the first animal to orbit Earth, we may simply find ourselves on a spaceship that can no longer support us.

Setting eye on Earth for the first time helped strengthen the nascent environmental movement, a kind of societal immune response to industrial effluents, off-gases, and other depredations of the modern world unleashed by the abundant energy stored in fossilized sunshine. The picture, later known as *Earthrise*, humbled those who saw it, a scrim of white clouds obscured even the greatest works of humanity. The nature photographer Galen Rowell called it "the most influential environmental photograph ever taken."

When NASA released *Earthrise* in 1968, someone had reoriented the image to make it appear as if Earth rose over the horizon of the moon, as the moon and sun commonly rise over the horizon when viewed from Earth. But that is not how it works in lunar orbit. Earth actually snuck around the left side of the moon, as viewed from the orbiting spaceship, and the astronauts saw the lunar horizon as a vertical fact and Earth just hovering in the void.

Humanity often seems to be in the habit of such misperceiving, changing the picture to suit our preconceptions. The close view is that Earth is simply the ground beneath our feet (though it is mostly water). Back on Earth, 1968 saw riots, protests, even war, as well as assassinations. Martin Luther King Jr. and Robert F. Kennedy died and Soviet tanks crushed the Prague Spring. At the same time, massive dams captured the flow of rivers from British Columbia to Nigeria. Luis Alvarez, who helped build the first nuclear weapon, won the Nobel Prize for his work in particle physics, and France tested its first atom bomb. Charlton Heston damned us as fools at the end of *Planet of the Apes* for the use of such weapons of apocalypse and Saddam Hussein seized power in Iraq. A military dictatorship took firmer control of Brazil, launching a wave of deforestation that would soon be visible from space.

By 1968, a new era in the history of the planet was also well under way—the Anthropocene, a union of what was once considered, perhaps foolishly, solely human and solely nature.

This new epoch remains ill defined. It may be a scant 250 years old, dating to James Watt's invention of a practical coal-burning steam engine. Or it could be as much as 50,000 years old, when humans began killing off other large animals in earnest, starting with our fellow hominids, the Neanderthals. Humans really got good at leaving our mark in the 1940s with the invention of nuclear bombs, which spread a unique human signature of rare elements like plutonium across the globe and marked the beginning of an ability to be the destroyer of worlds. Still, whatever date is chosen, it's a rounding error in the 4.5-billion-year history of this planet.

Through ingenuity and technology, *Homo sapiens* has become the dominant species on the planet. From deep beneath the ground, where we busily hollow out yawning cavities in pursuit of fossil fuels, to the skies, where carbon dioxide molecules released by our incessant burning will trap heat for longer than our species has walked the planet, we now drive the processes some still like to think of as nature. And just as a climate shift marked the turning point in the two most recent epochs—the cold, wintry Pleistocene to the long summer of the Holocene when civilization was born and flourished—so too our human-made climate change merits a new epochal designation.

The marks we have made are deep and pervasive. Humanity is writing a new chapter in Earth's history, and it may not remain a rounding error for long. Our mark lies in the rock being formed beneath our feet, in the chemistry of the oceans, in the composition of the air we breathe, even in the very evolution of life itself. Agriculture, cities, and sprawl drive the sixth or maybe seventh mass extinction in the planet's history. The choices made this century will help set the course of the entire planet for at least tens of thousands of years. If people, plants, and animals don't like the climate of 2100, 2500, or perhaps even 25000, they will have us to blame. And if human civilization is to persist to see that climate, we need to get busy inventing a better future. We have rules and religions to prevent murder, but far too few moral codes or stories to guide us as we attempt to survive and thrive on this planet for millennia more.

"We" is not the perfect pronoun for all this, of course. Not all of us have benefited from this world-changing transformation, or at least not equally. The luckiest few have grown fat as the primary beneficiaries of the environmentalist's paradox, which, put most simply, is: If the world is going to hell, why are so many people better off than at any other point in history? The middle-class "we" is now more than a billion strong—and growing. We have created technologies and systems that any individual can barely see let alone comprehend, many of which continue the seemingly implacable work of uprooting forests, besmirching the skies, and befouling the seas, with results that are all too visible. As a result, a creeping sense of doom and impotence pervades many of the minds of the best-off people of all time. Meanwhile, the richest among us, the almost all-male cohort who put the *anthropos* in Anthropocene, who mostly grew wealthy through uninhibited exploitation of the riches of this Earth, avoid the most numerous victims of this pollution, most of whom live in the future.

But people are also the first life to transform this planet with the possibility of consciously recognizing that world changing. Civilization has terraformed Terra herself, mostly unwittingly, just like the cyanobacteria that laced the planet's skies with oxygen long before us. Those unconscious life-forms evolved better ways to organize—plankton, kelp, ferns, trees—systems that last to this day. We call our evolved organizational structures farms, houses, and cities, but we also don't pollute only with one gas. We don't have to hit the world like an asteroid, wreaking blind havoc as we have too often done to date. We can be a force for good. We still have a choice—and time to make it.

Mother Nature seems to be losing the battle to live with her most innovative child. It's now up to us to grow up and control ourselves, recognize that we set the terms under which the rest of life on Earth thrives or dies. Were we to perish like the dinosaurs, only those animals and plants that survived our current reign of malign neglect would be around to fill the ecological niches left empty by our depredations.

And it's not just biology. Human dams hold back five times more water than flows through all the rivers and streams on all the continents. Human-made canals function as new rivers while natural rivers are channelized into canals. Reefs and atolls become full-fledged islands with sand imported from elsewhere or sucked up from the seafloor. Some 500 million metric tons of aluminum have been pulled from Earth's crust— enough to cover the United States in foil—and used to make everything from airplanes to beer cans. We have even made 6 billion metric tons of plastic, enough to cover the whole world in a thin film, in just the last fifty years or so. These are no minor perturbations. We have a global influence without any sense of global responsibility, a kind of haphazard world-changing impact, self-assembled from the activities of 7 billion- plus individuals trying to build a decent life, and a better life for their descendants.

As the crew of Apollo 8 showed, we will leave our mark on the uni- verse too, radio and television signals propagating into the void along with spacecraft, potentially bearing stowaway microbes to seed new life. If the *Curiosity* rover on Mars strikes water, Earth microbes contaminat- ing its instruments may find a new home. We have even begun to litter nearby planets—as well as Earth's orbit—with space junk. The astronauts who landed on the moon after Anders and his colleagues showed the way left an enduring legacy: trash—ranging from bodily waste to golf clubs. Will anyone go back and pick them up?

The Anthropocene no longer needs humanity to make its mark, though we make that mark more indelible with each passing day. This new age is not just climate change, it is everything change: the sky, the sea, the land, the rocks, life itself. Even if humanity were whisked away tomor- row in some kind of rapture, these changes would continue to propagate through time. If we want to ensure that this epoch is more than a blip in the history of the planet, it will take all of our smarts, sociability, and technology to begin to stop hindering the functioning of formerly self- propelled systems and perhaps improve them, such as balancing the flow of carbon into and out of the atmosphere.

As you will read in the following pages, that work has already begun.

Starting with the changes wrought in the global ocean and travel- ing underground, over land, into the sky, and finally back into space on potential journeys even farther than Apollo 8, I will introduce you to some of the men and women attempting to wrestle with our outsized role on the planet and what might be done to ensure that this human era endures. I will argue that we must accept the responsibility that comes with the power that humanity now possesses. We must begin to manage the planet, slowly, carefully, with room for both error and improvement, but inexorably.

This is not a settled debate. No one, least of all me, can yet say what the correct role for humanity is in this novel age. But it is certainly not preserving Nature with a capital N, somehow separate from humanity, or bringing stasis to what is and always has been the dynamic flux of life, chemistry, and plate tectonics. This is about managing change, adapting to it, and increasing the resilience of our civilization at the same time as we make more room for our fellow travelers on this life-bearing spaceship.

Human life in this new era is fraught with anxiety. Are there too many of us? Is the food we eat destroying us and the planet? Is our very own technology transforming us into mindless, atomized automatons? And what exactly is the impact of all that technology on the planet itself, like the jet contrails that scar the sky day after day? In the competitive thrust of modern life, what is lost? To me, the most important question is: Are we even ready to attempt to manage the planet better?

The stakes could not be higher. As the astronomer Carl Sagan famously noted, everyone who has ever lived, all the knowledge ever known, everything anyone has ever loved lives only on this incredibly small planet in the vastness of space.

We are now stewards. As the psalm goes: "The highest heavens belong to the Lord, but the Earth he has given to mankind." That stewardship may derive from the sty and the ward, as in warden of the pigsty—a messy, ravaged environment, full of muck, microbes, and only two kinds of megafauna: us and the pigs we raise to eat. Or we can be the kind of stewards that derive from the Old English word *stig*—house—full of many interlocking parts and functions that must be coordinated. We can make Earth a true home, the same kind of home we have shared with other animals and plants for the last 200,000 years—or an even better one.

What we stand to gain is nothing less than an enduring civilization and a firmer understanding of our planet and ourselves. We have arrived at a new geologic epoch of our own making. This can be very bad or even good. The choices made by our growing human family will determine that future. Only one thing is for sure: You can enter the Anthropocene any time you like, but you can never leave.

I argue the goal therefore must be to make an enduring Anthropocene, an epoch that, in geologic and civilizational terms, stretches into an era. This epoch could herald a new age of human adaptation, the greatest strength of a species that has shown an ability to thrive almost anywhere on the face of this planet.

This is a hard story to tell because it has no definitive beginning or end, and I could have gone anywhere, seen anything to tell it. We are all lost in the middle of it. That's why I started with the moment when we first got a more holistic glimpse of the problem from space, though the problem starts within each and every one of us. That's also why I'll start this journey of discovery in the largest realm of this planet, a place that seems beyond human control, or even, it sometimes seems, understanding: the oceans that cover more than 70 percent of the globe.

Our impact on the ocean will prove that the Anthropocene is now, as well as what this epoch might look like. We live these days on a science-fiction planet. I will then proceed to lay out the case for this new geologic epoch as it is written in stone. This epoch's existence argued for, I'll move on to examine other ways—what mix of plants and animals, wildlands and cities cover the surface of Earth—that will provide a metric for the job we will do as a world-changing species.

Finally, I will turn to the future, and how to make a better one, or at least a less destructive path on which to muddle through the next century and beyond. Now our every action has an impact, whether swaths of land sacrificed to produce energy or the encroaching fields of agriculture burning down rainforest remnants. What we lack currently is control, of both ourselves and the planetary systems we change in ignorance. This is not the end of the world. This is just the end of the world as we have known it.

Welcome to the Anthropocene.

Part I

Alter Earth

Chapter 1

Iron Rules

We take things out of the ocean we've never seen before and put things into it
we don't want to see again.
—MARINE BIOLOGIST CALLUM ROBERTS,
WINNER OF THE ROYAL SOCIETY WINTON PRIZE FOR SCIENCE BOOKS, 2013

The Southern Ocean is a forbidding place, mostly unvisited by humans. Winds of more than 120 kilometers per hour wreak havoc, and waves routinely wash up and over the bow of any ship that dares to venture so far south, drenching the decks in ice-cold sea spray that instantly freezes. Worse, rare freak waves can build to towering heights and flip a ship in place. Looming icebergs can fatally wound an unwary or unarmored vessel. Beyond this storm-tossed sea lies Antarctica, locked in deep freeze by a current that reaches 4000 meters from sea surface to ocean bottom and spans thousands of kilometers in its endless encircling moat. The Southern Ocean is where the mustachioed marine biologist Victor Smetacek, like any good scientist with a plan to change the world, hopes to change the sky with the power of the sea.

Victor's sailor father, Fritz, fled the German-speaking Sudetenland region of the former Czechoslovakia and, as a dedicated Communist, ultimately the Nazis. He washed up in India in 1939, where he dropped the z from Smetaczek. Among the teeming masses of Calcutta, he fell in love with Shaheda Ahad, a young woman from Orissa. Fritz and Shaheda moved to a onetime British hill station hundreds of kilometers to the west, closer to New Delhi. Victor, their youngest son, and his siblings were raised on a tea plantation 2000 meters up in the Himalayan foothills, far from any sea (though the hill station does surround a large lake) and just west of Nepal. Shaheda gifted her children, especially Victor, with charm and a lust for life. Young Victor indulged his passions for bird-watching, collecting butterflies, and hunting for food, starting to stalk jungle prey with his father's shotgun at the age of ten in the company of local farmers. He eventually graduated to stalking and shooting cow-killing leopards and even, once, an elephant.

Food remains a concern in India and, as a young man in the 1960s, the tragic famines of Victor's homeland drove him to sea. An article from a 1954 issue of *Collier's* magazine called "Bread from the Sea" inspired that flight in part, and in part perhaps to echo the journeys of his beloved father. In that article, the journalist Bill Davidson suggested that so-called phytoplankton—microscopic floating life powered by photosynthesis, often also called blue-green algae or cyanobacteria—might become a stable source of food for humanity. Smetacek used his Czech surname to land a scholarship to study marine biology at the University of Kiel in Germany in 1964. He had to learn German in a hurry from reading newspapers to do it, since his father had checked the "fluent" box on his application although, in fact, young Victor spoke fluent Hindi and little Deutsche.

At Kiel, Smetacek discovered the numerous organisms hiding behind the scientific term "plankton." This microscopic life fills the sea from top to bottom, extending even into the sediments of the seafloor. Since the 1980s, Smetacek has ventured aboard the Alfred Wegener Institute for Polar and Marine Research's sturdy icebreaker *Polarstern* to study these plankton in their native habitat: the Southern Ocean. It is there that phytoplankton show their full diversity, locked in an endless struggle with zooplankton, the abundant grazers of the sea. One family, the diatoms, their elaborate shells made of the same element as sand—baroque twisting nets, thistles and thorns, even pod-shaped landing craft—guard the ocean-drifting cells as they busily turn carbon dioxide into food using the energy in sunlight. Like a jumble of children's blocks of all shapes and sizes, a sample of seawater turns up countless thousands of the most abundant life-form on the planet, and the base of the global food chain. This variety of shapes and patterns shames plants on land, showing more range in size than the difference between moss and redwood trees, though it is the difference between the microscopic and merely tiny. There are more plankton cells in the sea than the current count of stars in the entire known universe. And some scientists argue that phytoplankton can start an ice age, given enough nutrients.

Plankton and their ancestors were the first geoengineers—large-scale manipulators of the entire planet and its biological, geological, and chemical processes. Some 2.4 billion years ago, photosynthetic bacteria began to bubble out oxygen. By 1.7 billion years ago, oxygen made up 10 percent of the atmosphere—a massive change in the chemical composition of the air. For the first time, fire was possible, if there had been anything to burn. Slowly at first, but steadily, the atmosphere became dominated equally by geology and biology, off-gassing from volcanoes, and emanations from life combining to wreath the planet in a particular mix of gases.

Now the burning of untold eons' worth of this captured and fossilized sunshine—otherwise known as coal, oil, and natural gas—has changed that mix of gases again. Carbon dioxide piles up in the sky, swathing Earth in a slightly different mix of opaque gases that prevent the planet from shedding back to space some of the sun's ceaseless heat. To throw off that greenhouse gas blanket, why not simply copy nature? Use photosynthesis to bind that excess CO_2 in carbohydrates and then bury that food beneath land or sea. Given that the plankton of the sea produce 70 percent of the oxygen in the atmosphere, these prolific tiny plants seemed a good place to start. Scientific experiments and even accidents of industry have shown that plankton blooms can be reliably induced with the addition of extra trace nutrients like nitrogen, phosphorus, and, in the case of Smetacek's work, iron. That's Smetacek's grand scheme: It shows how human intention may tinker with living flows—and even harness them. This is the Anthropocene as direct action, turning the wildest wilderness into a kind of farm, or a living system that serves human needs.

Smetacek and a boatload of forty-nine Indian and German scientists steamed out of Cape Town on January 7, 2009, bearing 20 metric tons of iron sulfate borrowed from a titanium smelter in Bremerhaven, Germany. He is too European now to be fully Indian and too Indian to be German, which made him the perfect person to pull together this joint expedition fueled by the redolent smells of curry. Plans hatched at a restaurant in Bremerhaven back in 2004 to fertilize roughly 300 square kilometers of ocean and to watch the results had finally come to fruition. But unbeknownst to Smetacek and his crew, this little iron experiment also had seeded the beginnings of a storm of controversy—and that storm was headed straight for the RV *Polarstern*.

Water covers more than two-thirds of Earth, yet the oceans often remain an afterthought for the land mammals currently at the top of the food chain. As Smetacek wrote in his final report from that 2009 cruise: "We live on a wet planet but, being land animals, we call it Earth." The world is really made by two great fluids—the ocean and the atmosphere—roiling against each other. Land merely impedes or redirects this ceaseless flow.

Personally, I fear the ocean, having grown up in the center of the continent known as North America and endured a perhaps too early exposure to the film *Jaws*. In my view, our distant fishy ancestors showed great wisdom in forsaking this dismal soup for the relative safety of land. The sea stews with danger, sharks and undiscovered viruses, all with a soupçon of human additions, like plastic ghost nets that snare the unwary. The oceans encompass the majority of the globe, so what happens to the oceans will determine the future course of the planet and the prospects for this current epoch.

Today, plastic litters the shores, not left directly by careless tourists but washed up from the sea, the last great garbage dump. Reefs that once abounded with life are coated in sediment and algae, showing glimpses of bone white and haunted by only the smallest fish, fit only to be boiled to add flavor to water. The giant snapper of yesteryear is no more. Signs of human activity have even reached into the bottom muck, where life proceeds at its slowest known pace and the future of rocks is laid down.

Mostly, the oceans reveal the depths of human ignorance. We have visited the bottom of the Marianas Trench, the deepest part of the sea, only twice—and one of those visits was by a Hollywood director. We have better maps of the moon than of the ocean floor in all its vast expanse, the first such global map delivered only in 1977—eight years after Neil Armstrong walked on the lunar surface. The life habits of eels and the origins of the Sargasso Sea remain a mystery, and the most abundant photosynthetic organism in the ocean wasn't even identified by scientists until the 1980s because the nets used to survey the sea had holes too big to capture a microbe that boasts 1 million cells or so per liter of ocean water. Marine scientists still routinely discover new species, even ones as large as dolphins. Ships on the surface can even go missing: The 100-meter-long, 1400-metric-ton Russian cruise ship the *Lyubov Orlova* disappeared in the North Atlantic in 2013; it still has not been found. This alien world exists in parallel to the continental crust we comfortably inhabit, refusing to forgive or forget by absorbing much of the extra heat trapped by accumulating greenhouse gases—even the carbon dioxide itself—and unleashing that heat to fuel extreme weather. The ocean's vastness defies human scope.

We don't even know enough to know what we're missing. This is the phenomenon scientists have dubbed "shifting baselines," or the fact that whatever one encountered (or didn't encounter) as a child is likely to be the basis by which one judges what is normal (or not) when it comes to the oceans or any other part of the nonhuman world. If, like me, you have never seen a two-meter-long grouper, then you are unlikely to think that its absence is unusual. But someone who snorkeled off Jamaica just forty years ago would note this dramatic lack. If right whales have not been seen off the coast of Europe in your lifetime, or the lifetime of even your great-great-grandmother, you are unlikely to think of them as missing— or, for that matter, think of them at all. It's a collective social amnesia that allows for continual, gradual decline.

In large part, we live with a sea haunted by ghosts. Roughly three-quarters of all large animals in the sea—fish, mammal, or otherwise—are gone. The ghosts can be dimly perceived in change. Revenants explain collapsed ecosystems like the cod banks off Newfoundland and even the price of lobster. It's as if we were trying to study the ecology of the Amazon rain forest based on the cattle and pasture that remain after the forest is gone. Everything is different. The bones of the Hawaiian petrel stretching back 1,000 years show a reliance on big fish as food while those from the last 100 years show a shift to smaller fish, squid, and the like—a new epoch recorded even in the hollow skeletons of seabirds.

What we know for sure is equally troubling. Overfishing has allowed jellyfish and squid to take over in many regions, devolving the oceanic food chain and returning the jellies to a dominance last enjoyed more than 550 million years ago. Meanwhile, the big predators of the sea, like the mighty bluefin tuna, keep getting smaller, no longer the 4-meter-long, 700-kilogram giants that could dwarf a moose. We kill sea turtles and other jellyfish eaters in large numbers, even if mostly by accident now with our fishing gear. The only hope for change may be that humans develop a taste for these newly dominant species like the jellies or lionfish and overfish their numbers back down, though this is not a perfect solution.

In just one year—1998—humans helped kill off one-quarter of the world's coral. The Indian Ocean alone lost more than 70 percent of its reefs due to higher water temperatures, causing coral bleaching. Warmer waters cause corals to expel the algae that live with them and the reefs lose their color. The corals are not dead—not yet—but they are stressed and, when confronted with other challenges, such as water pollution or over-fishing, can easily succumb. Nearly twenty years later, only some of these reefs have recovered and the last few years have again seen unusually warm ocean waters and more mass bleaching around the world thanks to ongoing global warming. Dead reefs in turn imperil the homes of fish and other sea life that more than 1 billion people rely on for sustenance.

More than 40,000 square kilometers of ocean floor are trawled or dredged every day, creating a map of regular track marks in heavily fished seas off Europe and North America. No one knows what impact this has on life at the bottom because it is largely unstudied and unobserved, but we do know that the North Sea was once covered in oysters and is now covered in shifting sands. The seafloor has been flattened, as if plowed, and trawling is not the only way we blindly impact the deep. A tin-based coating applied to boats to keep them cleanly slicing through the water turned out to make all mollusks grow penises and, even after a ban, con-tinues to mess with the reproductive systems of sea cucumbers living on the seafloor. Nor do we have any sense of what life is like in the true abyss around 5000 meters below the continental shelves.

We are even creating entirely new ecosystems, such as the life clinging to the rafts of plastic pellets (called nurdles) and microbits swirling in the great gyres of the world's oceans. This "plastisphere" supports an array of microbes, from those that feast on the hydrocarbon chains in the plastic itself to opportunistic microbes that make food from sunlight or feed on those that do. Scientists have found thousands of microbes floating on individual pieces of plastic.

Some of the prelapsarian glory we will never get back in our lifetimes, including the great pods of thousands of whales that once graced the global sea, fertilizing the phytoplankton base of the food chain as they traveled by defecating. And the sea absorbs much of the fossil carbon released by the burning of coal, oil, and natural gas, rendering the salty waters more acidic. "The last thousand years have seen our influences gather like a green wave, scarcely perceptible at first but slowly lifting and steepening over centuries to burst across the globe in the last sixty years," writes the marine biologist Callum Roberts in *The Ocean of Life*.

Despite all this human-made pollution and all these human-made challenges transforming the ocean and creating the conditions for a new geologic epoch, what worries some people most is the dumping of iron in a small patch of the southernmost sea.

The storm started with an email. ETC Group (Action Group on Erosion, Technology and Concentration) sent Smetacek an inquiry about his Indian-German iron fertilization experiment. The small environmental outfit started off twenty-five or so years ago working on protecting the livelihoods of the rural poor, but over time it has branched out into the environmental and socioeconomic impacts of any new technology. To ETC and, in particular, its program director Jim Thomas, Smetacek's experiment amounted to geoengineering—a deliberate intervention in the planetary environment. While the United Kingdom's Royal Society, the world's oldest scientific body, also included the phrase "large-scale" in their definition of geoengineering—and Smetacek did not plan to fertilize much of the ocean—Thomas and his peers felt that it was close enough, particularly in light of previous efforts to fertilize the ocean near the Galápagos proposed by a commercial company.

The mischievous Smetacek responded to the email on the first day out, noting that his experiment had "the explicit permission of the German and Indian governments, including the German Ministry of the Environment, which hosted the [Convention on Biological Diversity] meeting last May." That convention, often shortened to CBD by those wonks who traffic in its arcana, asked governments to refrain from such iron fertilization experiments "until there is an adequate scientific basis on which to justify such activities." Armed with that wording, Thomas took Smetacek's email to the German environmental ministry, at the time headed by the former high school teacher Sigmar Gabriel.

Gabriel professed no knowledge of the experiment and sent a request to his colleague, the minister in charge of research, Annette Schavan, that Smetacek's experiment be stopped immediately. This apparent contradiction within the German government spawned an international media frenzy and, by day three of the *Polarstern's* voyage south, the ship had to stop, leaving the scientists on board in limbo.

Yet this experiment was nothing new.

Smetacek had been chief scientist on a similar iron fertilization cruise back in 2004. What had changed? Just one thing: A peripatetic citizen-scientist named Russ George had attempted to make money from the idea.

A bearded, bespectacled, pudgy little bear of a man who looks like he just stumbled out of his hippie days and is prone to nervously biting his fingernails, George describes himself as a forest ecologist. He planted trees in Canada's forests to replace those chain-sawed to make way for roads. He even turned tree planting into a business of sorts, starting a company called KlimaFa (Climate Trees) in Hungary in an attempt to get paid to plant trees as part of efforts to draw down the CO_2 put into the atmosphere by fossil fuel burning. He planned to donate some of those forestry carbon offsets to cover the climate-changing pollution of the Vatican, that is, to make the Holy See carbon neutral. But now the bizarro entrepreneur, who also previously tried to sell cold-fusion devices (which have yet to be proven to work), had a new scheme: plankton blooms for carbon credits. In 2002, George chartered rock star Neil Young's wooden schooner *The Ragland* to test out the theory behind iron fertilization off Hawaii.

Encouraged by the results and interest, by 2007 George had formed Planktos Inc. and raised funds for an expedition to the Galápagos Islands to dump iron in the water. He proposed to make that expedition a commercial venture, much like his tree-planting scheme: He would sell the rights to the CO_2 sucked in by any subsequent plankton bloom. This is the core idea of an "offset": You reduce CO_2 by, say, fertilizing a plankton bloom and then sell the credit for that sucked-up CO_2 to someone emitting a lot of CO_2, like the owner of a coal-fired power plant. And Planktos wasn't alone in trying to sell the idea: San Francisco–based Climos had a similar plan, as did Virginia-based GreenSea Venture. The schemes attracted investors including Elon Musk, wunderkind of electric cars and space travel these days but not shy of geoengineering then or now. Nor was it confined to the United States. Australia's Ocean Nourishment Corp. floated similar plans. The scheme promised a forestry of the sea, but it really was more like boosting sapling growth by dumping a ton of fertilizer on otherwise degraded land in the hopes that something green would grow faster and more abundantly.

George's expedition actually set sail, embarking in the *Weatherbird II* on a self-declared "Voyage of Recovery" on November 5, 2007, with plans to fertilize an area of ocean twice as big as Rhode Island off the Galápagos. The old fishing boat bore a snazzy blue-and-white paint job, along with PLANKTOS in big capital letters near the bow, just behind the much smaller RV *Weatherbird II*. All that remained for the captain and merry crew of fresh-faced young environmentalists to do was find the iron to do it. The U.S. Environmental Protection Agency had warned George that loading up with iron from the United States and dumping it might constitute a crime, and that sent the controversial ship in search of friendlier shores, including a stop in Bermuda and an attempt to land in the Canary Islands to pick up some iron and, potentially, fertilize the Atlantic instead. All the while, Planktos sold carbon offsets on its website at the low, low price of $5 per metric ton.

At the same time, an international consortium of environmentalists mobilized to stop George's bid, including the captain's former colleagues at Greenpeace. The aggressive Sea Shepherd Conservation Society, which does not fear to tangle with Japanese whaling ships in the Southern Ocean, threatened to do whatever was necessary to stop such ocean pollution. Within only a few months, George went from testifying before Congress on the benefits of iron fertilization and carbon credit sales to being an international pariah. By December, the *Weatherbird II* had become a modern *Flying Dutchman*, homeless until the ship's crew promised to suspend operations and docked in Madeira. By February 2008, the Voyage of Recovery had been indefinitely postponed. But George's machinations for "organic gardening" of the sea and the doomed voyage of the peripatetic *Weatherbird II* did have a lasting effect. The CBD meeting in Bonn in 2008 explicitly took up this idea of iron fertilization, resulting in that request for restraint.

The convention states: "Such studies should be authorized only if justified by the need to gather specific scientific data, should also be subject to a thorough prior assessment of the potential impacts of the research studies on the marine environment, should be strictly controlled and should not be used for generating and selling carbon offsets or for any other commercial purposes." That's exactly what Smetacek had set out to do because, frankly, no one knew whether such iron fertilization would even work to bury CO_2.

Could plankton farming via iron fertilization poison the sea? Part of the reason for the storm of controversy also had to do with unknown side effects, perhaps toxic algal blooms, like the kind that kill sea lions and other animals with domoic acid off the coast of California. "This is crazy," Smetacek protests when I visit him in Bremerhaven years later, noting that the zooplankton known as copepods appear not to be affected by this poison. "All plants are toxic. The plants in my garden are toxic, but the deer don't notice."

Perhaps such fertilization could unleash a burst of the potent greenhouse gas methane? Or maybe a bloom could create a mobile "dead zone," like the ones at the mouth of the Mississippi caused by all the fertilizer washing off midwestern cornfields? That nitrogen works equally well for algae and, when the algae die, other microbes feast on the corpses, in the process consuming all the oxygen in surrounding waters. Anything that cannot move dies, and the dead zone recurs each summer, sometimes becoming as big as the state of Connecticut, or bigger.

The worst impact might be on the clouds and rain. In yet more proof of geoengineering ways, plankton help form clouds by pumping out gases like dimethyl sulfide—the compound responsible for the smell of the sea—while also covering the surface waters in a profusion of cells. Wind and waves loft these gases and cells into the sky. The particles from the ocean then serve as seeds in the sky around which water vapor can condense, forming the droplets that make up clouds. These clouds then drive rainfall around the globe. Some climate models suggest that fertilizing the ocean to make more microbes could end up creating less rainfall in Europe and South Asia. That seems worrisome until one realizes that we are already changing weather patterns with all our coal burning as well as the controlled burn of bunker oil to power ships. The leftover soot and ash from these controlled fires dust the sea with iron and other elements—and make for more clouds.

The surest sign that we have entered a new geologic epoch is the transformation of the ocean by us. The seas will not be the same in one hundred years: more acidic; probably fewer fish and krill; different shorelines as the waters rise. These waters will also, to some extent, be tamed for the first time in human history, seawater farmed for fish, algae, even perhaps electricity generated from harnessing the temperature difference between the top and bottom of the ocean.

Such changes by life are, of course, old, old news. The first such transformation of the planet began with just one. One tiny cell that had built inside itself a new pigment, a brilliant green, thanks to its ability to absorb only certain colors in the light of a younger, weaker sun. The pigment—dubbed chlorophyll by animals who rely on this one cell's innumerable descendants to power naming brains—channeled the energy in sunshine into splitting the surrounding waters of Earth's early oceans. Like its own ancestors, this blue-green cell then knit together the resulting hydrogen with carbon captured from the sky and sea to make food.

The new way of life prospered. From that one cell, millions bloomed. Together, they turned Earth's waters a murky green some 3 billion years ago.

The planet back then was a hot, violent place, churned by continuous eruptions of magma, wreathed in an atmosphere of "sewer gas"—a pungent and poisonous combo of ammonia, suffocating carbon monoxide, hydrogen sulfide, and the simplest hydrocarbon (the one used as "natural gas" fuel today): methane. But the tiny blue-green cells went about their life's work, over time organizing into thin mats of cells. These mats of bacteria busily turned sunlight into food. Year after year, the mats grew, layer by layer, forming small mounds or even columns. The abundance of life grew.

These columns provided the phalanxes of a budding geoengineering coup, although one that would take hundreds of millions more years to triumph. Because bubbling out of each column was a gas inimical to life on this Archaean Earth—oxygen.

Sulfurous microbial overlords ruled our planet's early history, stealing energy from the abundant sewer gas and sulfur-based minerals. As long as sulfur remained plentiful, their dominion was assured, or so the story scientists tell goes. But over eons, these early organisms ate their future—levels of sewer gas declined and venting volcanoes no longer belched as much of the sulfurous gas. With scarcity came innovation, and evolution created the cyanobacteria—tiny blue-green cells—that could use the old, easy way of making a living along with a new, harder but more abundant, lifestyle—using sunlight to split the copious water on the planet; H_2O turned to H_2 and O_2 via photosynthesis. The limits to growth had yet to be found.

The sulfur lovers found the by-product of this new way of life poisonous, dying when exposed to even minuscule amounts of oxygen. But the iron dissolved in the world's oceans seems to have protected the planet's first dominant life-forms for untold eons. The oxygen produced by the would-be usurpers rusted that free-floating iron, dropping it to the bottom of the ocean floor and, ultimately, created the red rocks visible today all over the world. Over a billion years, as oxygen began to bubble out of the oceans, the sulfur lovers were slowly driven into the dark, secret places of Earth—dense muck that keeps oxygen out, deep underground, even, eventually, our stomachs.

An oxygen catastrophe likely drove much of the diversity of early life to extinction, a mass death caused by the new masters of the planet—photosynthetic bacteria. The blue-green cells even evolved the ability to pull nitrogen out of the air and turn it into the proteins, including chlorophyll, each cell needed to thrive. New life-forms swallowed some of the tiny cyanobacteria and domesticated them into the living power plants known as chloroplasts of more advanced life: algae, eventually plants. Everywhere, more and more oxygen entered the oceans and atmosphere.

By roughly 600 million years ago, there was enough oxygen—and plants to eat—to allow a new form of life to evolve: animals. Jellylike creatures, sponges, frond-shaped plant eaters fixed in place began to consume the fruits of the cyanobacteria's labors, using the abundant oxygen to burn the food and releasing carbon dioxide as waste. This symbiosis has governed life on Earth ever since. This symbiosis is even recorded in stone and that stone dated to what's called the Ediacaran period by a later species obsessed with naming. By 540 million years ago, a profusion of life had emerged: from the lowly clam and coral to the buglike trilobite.

The record of life on this planet is clear. A hundred million years after the explosion of diversity of life in the sea during the Cambrian, the first plants colonized the land and raised oxygen levels to as much as 35 percent of the surrounding air. As a result, giant ancestors of today's insects could thrive, like dragonflies with a wingspan larger than most modern birds and centipedes the size of modern alligators. As the millennia progressed, geologic processes and ever evolving life conspired to bring oxygen levels back down to the comfortable 21 percent of today. The only constant is periodic mass murders, and this endless tumult of birth, death, evolution, and extinction finally gave rise to a species that could challenge the long reign of the photosynthesizers—Homo sapiens.

The descendants of those first cyanobacteria have now ruled the planet for billions of years, diversifying to become algae, organizing into plants, reaching for the transformed skies on tree branches knit out of air. Their transformation of the atmosphere was terraforming in the best sense from our perspective, because it made the world just right for us more than 2 billion years later. And now we're going to change the chemical composition of the atmosphere again and wrest global power from the photosynthetic microbes that made life as we know it possible.

Smetacek is one of a growing band of scientists called, by some, geoengineers. There's the physicist David Keith, busy building machines to pull CO_2 from the air or seed the skies with a haze of cooling smoke, which will shield the planet from some of the light and heat from the sun. He warns that humanity better come up with a more compelling rationale for saving the Amazon rain forest than its service as the lungs of the world. Otherwise, he can just build a machine that performs the same function of filling freshly cleaned air with life-sustaining oxygen and eliminate the need for conservation.

Further back in time, the brilliantined chemist Bernard Vonnegut, brother of the famous novelist, helped discover the scientific key to rainmaking in the 1940s, writing up his work in *Scientific American* in 1952. Tiny hexagonal crystals of silver and iodide released as glinting smoke into clouds provided a structure to the water in clouds that inspired water to form crystals as well, and crystals of water are the key ingredient of rain and snow. He suggested that such cloud seeding might do more than make rain and also allow us "to initiate and direct storms and to exercise a considerable control over the atmospheric circulation on our planet."

His more famous brother Kurt co-opted the discovery in his novel *Cat's Cradle* for ice-nine, a substance that freezes the world's water on contact, and indeed, silver iodide finds prolific use today in a sometimes desperate bid to control the weather with what we might call ice-six, given its shape. The Soviets used such cloud seeding to keep rains laced with radioactivity from the Chernobyl nuclear accident falling on Belarus and points west rather than on Moscow. The Chinese employed batteries of silver iodide in a bid to ensure sunny weather for the 2008 Olympics and are pondering its use to clean the choking smog from its cities' air today. The U.S. Department of Homeland Security has even revived research into controlling hurricanes through similar methods.

Even drought-stricken farmers in the Horn of Africa have induced the Kenyan government to experiment with silver iodide cloud bombing.

Who is to blame for the weather, anyway? Despite Vonnegut's invention, we're not to the point of dialing up a thunderstorm or shutting down a tornado. But we are at the point where, much like the flapping of a tiny butterfly's wings in that long-ago example of chaos theory, we are influencing outcomes in ways we can't foresee. If we aren't already, we will soon be to blame for the rain or shine. In a sense, the weather has become an artifact—a construct of human choices, like our choice to burn millions of metric tons of coal every day.

We still will not have achieved the ancient human dream of weather control that gave us deities such as Zeus, Thor, and the dragon Kuraokami or elaborate prayers and ceremonies for rain. But there has been a profound change in the weather. As the affable climatologist Stephen Schneider told a gathering of the American Academy for the Advancement of Science back in 1972: "Mark Twain had it backwards. Nowadays, everyone is doing something about the weather, but nobody is talking about it."

Smetacek does not think of himself as a geoengineer; that would imply too much control. Nor is he a seafarer in the grand tradition, with the face of a kindly German man ensconced in the brown skin of an Indian one. He doesn't go in for hazing at sea or other time-honored nautical rituals for first-timers crossing the equator or other, even more arbitrary, lines drawn on the globe. That just doesn't happen, "not on my ship," he explained to me when I visited him a few years after the research cruise in Bremerhaven—Germany's second-largest port and home of the Alfred Wegener Institute's fleet of research vessels.

Bombed nearly flat during World War II, Bremerhaven is both clean and industrial, an apparent contradiction only former Axis powers like Germany and Japan seem capable of achieving. The forest encroaches on Smetacek's home, which he shares with his wife of nearly forty years as well as the martens—weasel-like forest creatures that like to chew out the inner tubing of car engines. This patch of land is almost entirely of the Smetaceks' devising, planted with everything from bamboo and Sichuan pepper plants to the herb known as bog myrtle or gale that Vikings put in their beer to help them go berserk. But it was the quince tree in full flower back in the spring of 1986 that convinced Karen Smetacek to buy the house—with the addition of triple-paned windows to keep out the noise of cars rushing by on the Autobahn on the nearby embankment that is higher than the second floor of their home. Their warm partnership has endured despite Victor's frequent absences in the Antarctic.

Such adventuring seems appropriate for a man who spent his childhood splitting time between hunting and a trove of the *London Illustrated News* stretching back to 1880. "I was more steeped in Victorian culture than an Englishman," he observes with a chuckle. That may also help explain his graciousness and old-world charm. He is fond of a singsong string of "okay, okay, okay" or a breathy "ah" at every utterance of his guest, no matter how inane, before launching into his own explanation or theory.

He is an indefatigable host, picking me up in Bremen, most famous perhaps for its eponymous quartet of animal musicians: the outcast donkey, dog, cat, and rooster. This adventurer's party took on human thieves and won a home, according to the fairy tale, and they recur as a motif throughout the town, along with a smattering of "degenerate art" the Nazis allowed to survive as a warning and that the Allies failed to obliterate with bombs. But Bremen might be more justifiably famous as part of a powerful trading alliance that, arguably, laid the foundations for modern capitalism during the Middle Ages in Europe. Smetacek ascribes that to the lack of abundant local stone: no stones, no fortresses, which means no warlords and feudalism to enact merchant-crushing taxes. Teamed with Hamburg, Lübeck, and Danzig in the Hanseatic League that spanned the Baltic Sea, Bremen's merchant ships traded salted fish, coffee (a Bremen merchant invented the decaffeinated kind), and a host of other goods, setting the stage for modern capitalism as still practiced to this day.

Today, Bremerhaven is more of a boomtown for the offshore wind turbines that may make Germany's bold experiment to rely entirely on wind, solar, and other renewable energies possible. The country hopes to eliminate much of its climate-changing pollution, shifting away from burning the region's abundant coal to a reliance on sunshine and stiff Baltic winds to generate the electricity necessary for industry and other necessities of modern life.

I can attest to the stiffness of the Baltic wind, standing out at the edge of Bremerhaven. It is biting, worth bowing to as one takes in the kilometer after kilometer of cars waiting to board boxy car-carrying ships for export. A vertical wind turbine with its wavy white blades spins busily atop an old smokestack like an illusion of smoke. Beyond the tall earthen dike that shelters suburban German homes from storms like the hurricane-force winter storm that battered northern Europe in 2013, the wind cuts across the mudflats and grass and makes one more likely to seek hot tea than a mug of good but cold German beer. In the distance, massive platforms help the erection of the gargantuan offshore wind turbines. Let's hope our descendants will have better luck puzzling out their use than we do in understanding the megaliths that dot now mostly drowned Doggerland, the land bridge that connected Great Britain with Europe, which disappeared under the waves as the vast ice sheets of the Pleistocene melted. More land may be lost if we continue our fossil-fuel-burning ways, melting the vast ice sheets of Greenland and Antarctica.

Smetacek is also the absentminded professor, confounded by the winding medieval streets of old Bremen where he has lived for decades and mixing up the dates of a visit. He reminds me of some of the kindly but confused characters played by Armin Mueller-Stahl, a ubiquitous German character actor. In any case, a Smetacek-led tour always seems to end with food, whether a late lunch at the venerable Bremen Ratskeller (erected 1405), a quick trip to the market for dinner makings, or a pickled shrimp sandwich consumed while staring at mudflats.

Smetacek is unlikely to return to sea, he says while munching, though if anyone were to invite him, he would accept. He still teaches the occasional course at the Alfred Wegener Institute here in Bremerhaven, including How to Think, really a guide to critical thinking and creativity, in science or otherwise. Sample question: Which is heavier, wet air or dry? And, more particularly, why?* Smetacek has an effervescent mind, with bubbling interests ranging from archaic science embedded in Buddhism and Hinduism to what's really happening in the brain during sleep (he's sure it's "calibration," which in laypersons' terms is the body's sensory and motor systems getting back in sync). Sleep is a subject in which he has some expertise, given the inability to sleep on a ship rolling in the swells of the Southern Ocean, and thus rolling a would-be sleeping body back and forth in his or her bunk.

---

* The perhaps counterintuitive answer is dry air. Even though humid air feels heavier, water molecules are lighter than the clumps of nitrogen and oxygen that make up the bulk of the atmosphere. And, as Italian scientist Avogadro proved, equal volumes of gases at equal temperatures and pressures contain an equal number of molecules. The water vapor displaces the heavier N_2 and O_2, making wet air lighter.

His home office smells of smoke, and he calls it his cave, where he once plotted to turn patches of the ocean green. Smetacek's grandfather was a forester in the Sudetenland, so green is in his family's blood. A green clock on the wall of the Smetaceks' kitchen trills out different birdcalls on the hour, including that of the now rare blue warbler. A twilight tour of the sprawling garden requires tall rubber boots to ward off the ticks that spread brucellosis. He warns me to check myself carefully in the morning.

Smetacek takes me to the Wegener Institute as part of his grand tour, and the opening of the annual polar photography contest. It features pictures of people, places, wildlife and, most compellingly, the elements themselves from both poles, the photographers drawn from most nationalities with a permanent research presence in Antarctica. The institute focuses its research on the poles, just as its namesake was primarily a polar researcher, though he is remembered most for his now accepted, once ridiculed theory of continental drift down the long ages of geologic time as well as postulating the existence of the supercontinent Pangaea. But it was the elements in Greenland that did poor Alfred in. Six months after he disappeared on an expedition, his body was found buried, with his skis marking the gravesite. It isn't known who buried him, but his body remains there.

The institute owns the *Polarstern* (Polestar), a hulking ship that can overwinter in Antarctica, which allows research such as an underwater laboratory for sound in the least noisy ocean in the world (spoiler: still loud with crashing ice, especially when the *Polarstern's* powerful engines are about) or sea ice surveys. Krill, the shrimplike food of big baleen whales, remain a particular focus.

It is the krill that are responsible for the rusty red trails that mark the preferred path of penguins, as pictured in one of the contest entrants' photos. Early Antarctic explorers spoke of seas turned red by an abundance of the tiny animals. "Krill is like a sponge," Smetacek enthuses, "a reservoir of iron." The ubiquitous whales of the Southern Ocean used to keep that iron in circulation, Smetacek is sure. Whale poop fertilized the oceans like manure from cows and buffalo on land, only the whale's liquid poop disperses in the water column, perfect to feed a plankton bloom and keep the cycle of life cycling. Here sperm whales play a key role—pooping out nutrients captured at depth and adding to the overall total, according to research Smetacek helped conduct. But people hunted out most whales decades ago, slowing this vital cycle.

"Give me a half tanker of iron, and I will give you an ice age," the oceanographer John Martin growled during a lecture at the Woods Hole Oceanographic Institution in 1988. That's because the tiny plants of the sea require one nanogram of iron per metric ton of water to grow, or roughly a paper clip's worth in several hundred thousand Olympic-size swimming pools. But that's not why Smetacek initially dismissed the idea. He surmised in an article in the German newsmagazine *Der Spiegel* that plankton were more likely to move up and down in the water column, rather than sink and bury carbon with them. Smetacek says that Martin forever called him "the elevator guy."

Martin's other critics noted that maybe 10 percent or less of any fertilized bloom would actually reach the seafloor and bury the carbon with the corpse. Still, why not find out if the gruff, bearded oceanographer from Moss Landing Marine Laboratories in California was right?

One must travel to where the plankton still rule, almost uncontested, to truly understand this fundamental planetary cycle, combining biology, chemistry, and geology. But the Southern Ocean is, in the words of Smetacek, "one of the least attractive places for human beings on Earth," due to heaving seas, steady, strong winds, constant temperatures of roughly 4° Celsius in summer, enveloping fog, and gray socked-in skies. Smetacek and other scientists and sailors like him find themselves alone at sea, with neither fishing vessels, navies, nor tourists for company.

The Southern Ocean is unique, not least for being left out of the list of the seven seas. Its squalls and storms prevented circumnavigation of the globe for decades and remain a tremendous barrier to shipping, moderated somewhat by the great geoengineering project known as the Panama Canal. Gyres and eddies form at will in these storm-tossed waters that also boast some of the richest array of life on the planet—from tiny krill to the world's largest animal, the blue whale. The ocean is also one of the single largest sinks of carbon dioxide from the atmosphere. Upwelling deep waters that last saw the light of day in the Arctic seas between Greenland and Norway absorb CO_2 from the sky, enough to offset more than half the greenhouse gases emitted by the entire United States. These bottom waters also carry all the nutrients necessary for plankton to bloom, except iron, which rusts out quickly in water.

Decades of research cruises to study the oceanic food chain in the Southern Ocean prepared Smetacek for what may become his defining work: three cruises to fertilize plankton with iron and prove that the resulting blooms could bury carbon dioxide deep beneath the sea. With the help of these single-celled plants, Smetacek hoped to transform the atmosphere and, in the process, extend humanity's reach into the most powerful biological loop on the planet, one that we rely on to breathe. To do that, Smetacek transformed the *Polarstern* with its deep draft and ability to ride out overtopping waves from a nomadic hunter of ocean knowledge to a stationary farmer tending a watery garden and its planktonic bloom.

Using the iron sulfate produced as a waste product by a Kronos titanium production plant back in Bremerhaven—sold by the bag as a lawn treatment and available right in the recently dredged port—Smetacek and his colleagues planned to supply what the plankton lacked. In Greek mythology, Kronos was the father of Poseidon, god of the sea. Kronos ruled the universe until deposed by another of his sons: Zeus, the sky god. Now Kronos's iron would serve to rescue his sons' mythical domains—sea and sky—from the CO_2 released by all those pesky fires set by humanity. Smetacek would put a little piece of Germany afloat on the seawater to make the diatoms bloom.

The Southern Ocean stood athwart such plans, no easy place to do science even in a "luxury vessel" like the *Polarstern*, as Smetacek describes the research ship. Otherworldly green icebergs formed under floating ice shelves drift in this sea, a cup carefully placed on a table slides and crashes to the floor, or an invisible bully shoves the unwary scientist down as the ship heaves. It is possible to fly upstairs or find it nearly impossible to climb downstairs. The nearest land—Antarctica—remains the only continent not permanently inhabited by the world's most invasive species: *Homo sapiens.* But even Antarctica, the frozen continent, is pockmarked with signs of humanity, from research stations to melting glaciers—and the entire ocean has literally soured as a result of human fossil fuel burning and its attendant CO_2.

The Southern Ocean is less visited than other seas, but it is no less affected. It will get deeper and hotter as it melts ice and CO_2 piles up in the atmosphere. Whalers have decimated (or more) the population of cetaceans, some of which are now in the midst of a long, slow, or even impossible rebound. The krill are disappearing, perhaps due to the loss of the whales, perhaps climate change, perhaps overfishing, perhaps all these insults and more.

Bringing back such sea life is what brought Russ George back into geoengineering in 2012. The salesman pitched the wealthy libertarians known as "seasteaders," who want to build rules-free societies on the high seas free from nation-state interference, on iron fertilization as a business plan for their platforms. But the libertarians did not bite, this time.

So, far from the watchful eyes of the world's environmentalists, George partnered with the Haida people of British Columbia to dump iron at sea in a bid to restore the salmon. Anecdotally, volcanic eruptions in the region had reliably prompted record salmon runs, perhaps linked to the deposition of iron at sea and its trickle-up effects on the food chain.

George and a band of Haida fisherfolk set to sea in a rusting fishing vessel. The plan was to dust the oceans with iron, a couple different types, from the same fertilizer used by Smetacek to the same dust turned into steel in foundries. As the Haida crew dusted, a bloom of some sort seemed to appear a few days later in the satellite pictures from NASA. As word of the unexpected mini-geoengineering test leaked out, a global media frenzy kicked off anew. George, as rogue geoengineer, found himself hacking the planet, much to the rest of humanity's chagrin. Particularly since the stated desire of the Haida to rebuild salmon populations seemed to be at odds with George's oft-stated desire to bury carbon. Because if the carbon ends up in the salmon, then less carbon gets lost at sea and instead is recycled back into the atmosphere as the salmon get eaten or rot.

The salmon have indeed returned, though it's hard to say whether there is a connection between George's effort and the record pink salmon numbers up and down the West Coast of North America in recent years. It could be successful ecosystem restoration, or it could be coincidence. The sea is a complicated place, and the Haida aren't sure themselves, parting ways with George in the wake of the scandal. So was this an instance of villagers taking care of a problem in their backyard, as George argues, or a case of native peoples who long relied on a wild resource attempting to husband that resource back to health? Or was it rogue geoengineering?

If the latter, are the farmers of my native Midwest rogue geoengineers, given the annual dead zone in the Gulf of Mexico? Or are the cities of the world, the Detroits and Stockholms and Qingdaos, changing the planet by prompting "green tides" with sewage? Intention is all, apparently.

That and commercial prospects. As the AWI glaciologist Heinrich Miller told me over coffee at the polar photo exhibition: Geoengineering "cannot be a business option." He thought something similar to the Antarctic Treaty that has kept the icy continent (mostly) unspoiled might work for governance of geoengineering as well, or perhaps some kind of a global tax on CO_2 pollution that then funds a geoengineering mission administered by the United Nations. Then again, technocracy mixed with bureaucracy does not have the best reputation for delivering concrete results.

At the heart of this iron fertilization story is the idea of bringing back something that was lost, as well as cementing into place something new: a new stewardship of the oceans given humanity's world-changing impacts. But, given shifting baselines, how do people even know what is gone?

For that, travel to the middle of the Pacific—an ocean so vast that it takes two weeks just to steam across without sight of land—to an uninhabited atoll named Palmyra. If there is a pristine reef in the Pacific—and "pristine" is an awfully hard word to live up to—then Palmyra is it, the tip-top of a volcano ringed in 65 square kilometers of coral halfway between San Francisco and Auckland. This is the wild as left to its own devices, or as much as is possible in these latter days of humanity's world-straddling impacts.

The atoll was saved from life as a nuclear waste dump or rocket launch pad by $30 million from the Nature Conservancy and now doubles as a research outpost teaching us that coral reefs might just be top-heavy in the predator department. Contrary to scientists' expectations, sharks predominate on an undisturbed Pacific reef, free of most fishing pressures, pollution, or the problems of tourism. In fact, sharks make up 60 percent of all the fish on Palmyra's reefs. This means that, unlike other marine parks, the fish here do not grow huge or live long lives, living in a state of nature that could be described as nasty, fearful, and short. The same way of life can be found at the other remote atolls of the Pacific: the rest of the 1,500-mile-long chain of Line Islands, Kingman, the Phoenix Islands. Sharks rule when humans are largely absent.

Even here in Palmyra, humanity has made its mark, from a lagoon reshaped to suit the needs of warfare in the 1940s to invasive palms taking over from *Pisonia* beech trees and reducing the number of manta rays in the waters offshore. Even in this remote necklace of islands, plastic washes ashore and climate change slowly warms the waters. The tuna fishing is not what it used to be, as the ranges of such slow-growing species diminish.

Still, the relatively healthy reef here has proved more resilient to such challenges than peers elsewhere that face even more threats in combination. Palmyra has much to teach us about what a refugia from the coming changes might need to survive.

The poster child of ocean health has to be coral. The great reefs of the Caribbean bleached out, remnant skeletons of once vibrant ecosystems. The picturesque reefs of Hawaii coated in waving green slime (hello, algae!). Even the Great Barrier Reef is challenged by coastal runoff, human construction, and climate change. We need coral, not least for the billions of dollars (and uncountable sustenance) in fish such reefs provide. The key to ensuring an enduring people's epoch is not a return to some oceanic Eden but rather enough functioning reefs and other ocean ecosystems to provide the services we humans rely on: food, jobs, coastal protection, among others.

Unlike sharks that prey on the old, weak, and sick, humans pull out fish of all sizes—and waste most of the catch. As a result, fish are evolving to be smaller, to be ready to reproduce younger: people function as an evolutionary force. The reason for all this global change is simple: commerce on a scale never before seen.

In the end, commerce may be the only thing that motivates humans to determine whether iron fertilization even works. Replenishing all that missing krill—a vital resource for fishing countries like Norway and Japan—may be the route to broader acceptance of iron fertilization, Smetacek surmises.

The *Polarstern* itself and its six decks, several cranes, and 20,000 horsepower of thrust for smashing through ice is, of course, made of iron, and its passage sheds the vital nutrient. But it is the iron dumped in her propeller wash by Smetacek and his colleagues that really raises iron concentrations in the surrounding seas. Cut off from most continental dirt and dust, the plankton of the Southern Ocean cannot get enough iron to grow. Fertilizing these waters with the metal could promote blooms that then suck CO_2 out of the air. When the microscopic creatures die, they sink in tens of millions to the bottom of the ocean and bury the carbon with them. Such fertilization might ultimately even help whale populations rebound, abundance propagating from plankton to krill to blue whale.

But the trick—what Smetacek must do—is proving that iron fertilization can work to draw down CO_2. After all, ocean waters tend to mix. Smetacek's solution was to fertilize a self-contained swirl of water that can maintain its shape for weeks or even months. These eddies form when the fast currents of the Antarctic Circumpolar Current meander into loops that ultimately detach and continue to spin for a while until they slow and dissipate into the surrounding waters. Like a black hole, these eddies trap whatever floats into them for as long as they last. Thus, on January 21, 2004, *Polarstern* steamed its way from Cape Town, with Smetacek aboard as chief scientist, to a rendezvous with such invisible whorls in the latitudes known as the Roaring Forties and Furious Fifties for their powerful winds and waves.

Chunks of ice also ply these waters. One encountered on the way south rose 60 meters out of the water and spanned hundreds of meters in length. Shaped like a squat sphinx, its "head" balanced huge chunks of ice like an oversized crown and waves surging up its sides polished its flanks to a gleam. Even as Smetacek and his fellows watched the ice change color from pale blue to gleaming white with shades of gray in between, pieces of that crown began to tumble down under the warm sun. Before the scientists' eyes, the iceberg collapsed like an avalanche, splashing into the water below. Minutes later, only a field of ice rubble remained—a spectacular reminder of the polar meltdown humanity is currently creating through our excess greenhouse gas pollution.

The *Polarstern* hunted like a dolphin—with sonar—for its elusive prey, the telltale signs of an eddy, invisible to the human eye. February 13 dawned rough with heaving seas, but *Polarstern*'s crew had skillfully brought the ship to the center of such a sworl. Two scientists wore plastic coveralls and gas masks to mix the irritating iron sulfate powder into the seawater. Using a funnel, they dumped the resulting solution into the

Polarstern's tanks to be stirred. Then, circling out from a buoy placed in the eddy's center, Smetacek and his team started to spew 7 metric tons of the ferrous solution in the ship's propeller wash through a 5-centimeter-diameter hose, "plowing" a circular path in the eddy of some 167 square kilometers.

Adding 0.01 gram of iron per square meter, Smetacek dusted the ocean, creating levels similar to those found in the wake of a melting iceberg. Foam lay like dust on the water in the wake of the *Polarstern*. Smetacek and his colleagues settled in—dodging the occasional roaring storm—to monitor the fate of their bloom. "We have shifted gears by moving from the uncertainties of the hunter, full of apprehension as to what the next bend of the front will reveal," Smetacek wrote in a progress report, "to the fatalistic patience of the farmer, watching the crop develop in the painstakingly selected field." Only their field was in the midst of a frenzied sea, blasted by winds of at least 40 kilometers per hour on the best days. At least such conditions were good for mixing the iron into the ocean's waters.

Take the stairs when it's not windy out Smetacek advises as we head to the AWI cafeteria for lunch. When it's windy, the turbines rising offshore offset the electricity used by the elevator. Ralf Rochert, AWI's communications guru who handled the iron fertilization firestorm, has joined us. Generally speaking, the Germans don't support Smetacek's fertilization efforts, he explains, ascribing that to a long familiarity with mechanical fixes to complex systems. Such fixes always have side effects that then require further fixes with new side effects, in a seemingly endless effort. So the iron fertilization perturbs a food web that then must be restored or too much CO_2 is pulled from the atmosphere, setting off an ice age and a rush to burn as much coal as quickly as possible to warm things back up. Technocracy dies from complexity.

Scientific opposition is sometimes even more rabid. Ocean fertilization finished dead last in voting on geoengineering techniques at a symposium held at Cambridge University in 2010, judged more dangerous even than Keith's stratospheric haze. Smetacek describes one opponent as practically foaming at the mouth as he denounced the possibility, preferring instead his own work on chemically capturing CO_2 at power plants and burying it underground. Prominent scientists argue ocean fertilization can never work.

But we don't know how dangerous (or not) ocean fertilization is—and we won't know without experiments.

En route to pick up another guest at the train station, Smetacek and I stop off at the Devil's Bed, an ancient stone megalith from the last Ice Age, a remnant of the drowned culture of Doggerland. Prehistoric peoples laid out massive stones here, perhaps, Smetacek posits, as observatories to confirm that the days were not truly growing shorter or longer. Much like modern scientists and their experiments, these prehistoric astronomers might have perpetuated their observatories with a simple argument: If we stop watching the skies now, the weather might really get bad.

As the scientists watched on the 2004 research cruise, *Chaetoceros atlanticus*, *Corethron pennatum*, *Thalassiothrix antarcticus*, and nine other fantastically named species of diatoms grew in abundance down to depths of as much as 100 meters over the course of two weeks or so. By the middle of the third week after Smetacek stopped adding iron, measurements from deep casts with a rosette (a circular frame holding cylinders to capture seawater) showed the bloom beginning to die. So many diatoms died, in fact, that they overwhelmed any natural systems for decay and fell like snow below 500 meters in depth. At least half of those corpses sank below 3000 meters, and mats of the stuff, looking like wool made from glass, littered the seafloor.

In some sense, fertilizing the ocean with iron just replaces the krill and whales culled by human hunters, which is what the Anthropocene could become in its best guise. It is even natural: The waters nearer to Antarctica reliably turn pea-soup green at the end of the austral summer in November when the sea ice retreats.

Smetacek's experiment was a success, inducing carbon to fall thirty-four times faster than natural rates for nearly two weeks—the highest such rate ever observed for CO_2. Smetacek had created a conduit of plankton between an increasingly burdened atmosphere and the deep ocean. As he now says, "Unfortunately, [John] Martin passed away before I could tell him that he was right."

But the trick is keeping the carbon down there. For that, Smetacek's work showed that a bloom would have to be the right kind of diatom. A subsequent cruise found that iron fertilization could easily fail if these particular diatoms didn't bloom, or if other essential ingredients—like enough silicon for the diatoms to build shells—were also lacking. The pod-shaped, gelatinous, see-through animals known as salps can mess up the carbon burial too, eating the plankton instead of letting it drift to the bottom. Even the krill can hamper the carbon sequestration; recent observations have shown that krill sometimes dive all the way to the sea-floor, stirring up the sediment to feed and causing more of the would-be buried carbon to come back into circulation.

The ubiquitous grazers of the sea—the copepods and other zooplankton that have forced diatoms into the wide array of defensive shapes—stand ready to eat and respire anything scientists might care to bloom. What we know of copepods comes from studies in beakers, but Victor is sure the microscopic animals behave quite differently in the open ocean. Perhaps further iron fertilization experiments could be masked as a study of the biology of how these grazers control the numbers of phytoplankton.

This is all a transition, in thinking as much as in reality. The ocean is no longer a vast, unknowable, untamable wilderness, but a new frontier. A land animal has come to tame the sea, this heaving, alien world made legendary in myths as far back as Gilgamesh. This ancient force that once hid gods that had to be supplicated before any portion of it could be safely crossed is now the first arena for large-scale manipulation of the planetary environment, manipulating one of the two great fluids for the benefit of the other: the atmosphere. The sea is becoming a farm. Cages for rearing salmon crowd the waters from Scandinavia to Chile, and ponds for raising shrimp have replaced half the world's coastal mangrove swamps.

Smetacek doubts that someone else will pick up his work anytime soon, and do the real science necessary to further prove how iron fertilization could work. After all, another United Nations treaty—the London Convention on the Prevention of Marine Pollution—now explicitly forbids "any activity undertaken by humans with the principal intention of stimulating primary productivity in the oceans." The seas cannot be farmed this way, at least not without the backing of a major nation-state. "It would take too much footwork now," Smetacek says. "What a pity."

Iron fertilization, artificial islands, or any of Smetacek's numerous other ideas could happen, if someone is interested in making a push. But not him; he's not interested in making a fuss anymore, or so he says now.

It is through krill-restoration efforts that experiments with ocean fertilization will most likely return, Smetacek thinks, or perhaps an attempt to bring back the whales. In the meantime, other scientists do try to push the science forward, monitoring diatoms with new tools such as battery-powered autonomous robots that glide through the seas like mini-missiles taking in data, backed up by measurements from the thousands of Argo floats that now dot the sea. Natural blooms seem to bury at the bottom of the sea roughly 20 percent of the carbon dioxide the phytoplankton take in during photosynthesis.

The dream of ocean fertilization is deferred, if not dead. The nutrients carried to the sea by the Amazon River can create a bloom that stretches more than 6000 kilometers from South America to the coast of Africa, but humans cannot fertilize even an area the size of Manhattan. The oceans may not be deliberately fertilized in Smetacek's lifetime, or any other.

For his part, George has been working on a rock musical to muster more support for his efforts: *40 Million Salmon Can't Be Wrong*. The Haida have offered data from the 2012 cruise to scientists, free of charge, asking only for "co-authorship on any publications."

Although "data can always be faked," as Smetacek observes, one thing that cannot be faked is the record written in stone as the sediment of the seafloor piles up, slowly moving toward a collision with continental crust and future eons to be spent perhaps thrust up high into the sky, like the crumbling carapace of the Himalayas. Future geologists will be able to see what people today have done to the seafloor, and that will form the core of any geologic argument for a new epoch—the Anthropocene. We have now seen how humans are changing the world on the vastest scale imaginable—the oceanic, from the bottom of the depths to the deeps of Earth itself, where the distant past may reveal a map of the future, both near and far—and how those like Smetacek may make that change even greater.

Smetacek may not even have to do anything; the meltdown of Antarctica may do his work for him. As the glaciers dwindle, melting down as the globe warms under the blanket of greenhouse gases added by human pyromania, long-frozen iron fills the sea, fertilizing yet more blooms. Or humans could intervene more forcibly, farming algae or trees to dump in the deep sea. As the possibly apocryphal Chinese saying goes, "One generation plants the seeds, the next relaxes in the shade." Perhaps the purpose of this Anthropocene idea is to encourage us to plant seeds.

Smetacek still has that drive to make the world right, or at least help people to think clearly about the challenges we now face. Maybe it will take the threat of something like geoengineering to finally bring law to the seas. We are in a new age if some of us can contemplate intervening in the most powerful biological loop on the planet, one that has guided Earth for billions of years, through countless geologic epochs. Phytoplankton have nurtured the profusion of life on this planet, so now perhaps it's time for them to be nurtured. "There's no point in hanging on to things or saying it used to be like this. That's changed, anyway," he observes, with perhaps a tinge of regret in his voice. "We've changed everything."

Chapter 2

Written in Stone

The human race exaggerates everything:
its heroes, its enemies, its importance.
—CHARLES BUKOWSKI

Were it not for an unpleasant dean of fossil fish, Jan Zalasiewicz might never have come to the Anthropocene. That voyage started with grapto-lites, whose name means "written in rock." These tiny, branching animals fooled no less an authority than Carl Linnaeus, progenitor of the modern scientific protocol for naming species. He considered them artifacts, pic-tures thrown up by the natural working of the rock to fool the unwary rather than true fossils of ancient life, pseudofossils in old Carl's view.

But animals they were, living in colonies of tiny cuplike structures known as theca. The graptolites built many theca together to form beau-tiful branched structures, like today's corals, that drifted in ancient seas. When the colonies died, the graptolites fell to the ancient seafloors, untouched by light or other animal life, leaving them to be found in rocks of a certain vintage all the world over in this latter age.

Climate change killed almost all the graptolites, or so scientists cur-rently speculate. An epic ice age 440 million years ago helps mark the end of the period known to modern geologists as the Ordovician, named after an ancient Celtic tribe in northern Wales that resisted Roman rule. Just as cruel Agricola ordered the Ordovices wiped out, the world-girdling ice left only a few species of graptolites alive.

This geologically abrupt disappearance makes them the perfect fossil to mark the end of the Ordovician. So there is a golden spike hidden in a rock face outside the village of Moffat in Scotland that definitively ends the Ordovician, marked by the few surviving graptolites, in the Linnaean terminology, *Akidograptus ascensus* and *Parakidograptus acuminatus*. The marker is not an actual spike of gold but the line formed by a new layer of a darker shale. Wherever these two graptolites can be found in the rock, a geologist can be sure of the age of that rock. That's what makes them an index fossil.

It is index fossils that interest geologist Zalasiewicz, who is now in charge of the effort to determine whether the modern era deserves its own epoch: the Anthropocene, or recent age of man, and, if so, what its index fossil might be. He must convince the peculiar tribe of scientists known as stratigraphers—keepers of the geologic timescale—that humanity deserves its own place in the rock record, the only species to have an epoch named after it.

An index fossil defines an age—the presence of a certain graptolite can indicate the finest distinctions of a million years or so in the rock record. That's because graptolites had a habit of changing shape, spreading through those ancient oceans and then dying out as the climate or some other condition that affected the sea changed. That means successive layers of ancient rock hold successive species of graptolites recognizable by their unique forms, whether the blade of a jigsaw or the prongs of a tuning fork. Find a certain graptolite, and a stratigrapher can tell you exactly when that rock was formed, and Zalasiewicz became an expert in their branching forms instead of pursuing his first love: fossil fish. He even has black-and-white electron microscope pictures of two particularly fernlike specimens affixed to his office door, opposed side by side and meeting almost in the shape of a heart.

The modern era may have its own index fossils in so-called technofossils, or all the strange goods humanity now produces that might be preserved in the rock. Or so Zalasiewicz speculates, which is why I traveled to Leicester in England to visit him. He is one of the last of the generalist geologists, with interests ranging from graptolites to how cities press down on rock and reorder the layers beneath them. He is a tall, thin man with the rangy fitness of a field geologist paired with the formality of the son of refugees raised in deepest Lancashire. On the day we met in his office, he had just returned from student fieldwork in the only desert in

Europe. In other words, he looked tanner than the gray skies of England would normally permit.

Rocks fill his office—volcanic rocks veined with minerals or the perfect spiral of a nautilus fossil—and books, Simon Winchester's geologic thriller *Krakatoa* and James Lovelock's spiritualist science tome *Gaia* next to biographies of Thomas Henry Huxley and Charles Darwin and across from *Geology of Poland* and *Climate Change*. Students' work piles high on the tables, sometimes held in place by a colorful hard hat instead of a rock sample, along with back issues of *Eos*, many still in plastic. Several charts of deep time are tacked to the walls. His office looks out on a grassy slope dotted with wildflowers backed by a wall of trees that hides the university campus and town beyond. A fresh crop of students appears year after year, and Zalasiewicz helps them identify difficult rocks or, when their time has ended, find jobs in the oil and gas industry. The Anthropocene, if it exists, is no doubt made by these geologists, not just in naming or cataloging but in generations of rock lovers finding the coal, oil, and natural gas that, when burned, creates the putative epoch's most significant marker.

Zalasiewicz completed his scientific training "rather inefficiently," as he puts it, before joining the British Geological Survey. An interest in the practical application of geology led him to public outreach, and then his innate talents led to some popular writing, including four eminently readable books, including *The Earth After Us*, which speculates on what alien geologists might find—and, more important, identify—about humanity eons hence. In other words, will there be an Anthropocene in the record or not? As Zalasiewicz notes of our species, with his icy blue eyes giving off just a hint of bemusement in the crinkles around the edges, "We've reset the Earth's biology." Just like those first photosynthesizers poisoned their living ancestors.

That's part of what convinces Zalasiewicz that humans deserve an epoch, and what draws him into the generalized debate surrounding this resurrected idea. In the meantime, his day job requires playing What Will Last? with his students. In the case of Leicester, the answer may be nothing, or perhaps heaps and piles of stones that will be far out of place thanks to the cathedral, abbey, and other stone buildings; minerals and metals redistributed along rail lines and modern building sites; strange waterways carved out of the ground from the network of canals. All of this may be visible far into the future, just as the two-thousand-year-old Roman baths can still be seen today. These signs are built for deep time, but are they as widespread as the graptolites once were?

In windowless conference rooms, in the pages of august scientific journals, in the hallways of international meetings, and over boozy lunches, there is a fight going on, a scientific battle. It is an academic war over whether any such thing as a geologic epoch tied to humans exists and, if so, when exactly it started. At stake is the possible location—temporally, spatially—of a new golden spike, perhaps literal, perhaps not, driven into a section of rock or mud that typifies the geology.

As chair of the Anthropocene Working Group of the International Subcommission on Quaternary Stratigraphy, to give the full formal title, Zalasiewicz is perhaps the world's foremost specialist in Anthropocene geology. He became involved through the Geological Society of London, an august institution whose members have acquired the ancient and time-honored right to a free lunch with wine. The wine lends a certain conviviality to attempts to wrest a story from inert rocks. "It really is a couple of sandwiches and a glass of wine, so the singing doesn't usually break out," Zalasiewicz demurs. "It's just enough to keep things, well, you know . . ." And I think I do know.

At one such lunch in 2008, Zalasiewicz noted to his fellow esteemed geologists that this word—Anthropocene—was out there, being used, apparently unironically, by scientists (if not geologists) including the Nobel laureate and atmospheric chemist Paul Crutzen and the biologist Eugene Stoermer, who coined the modern iteration of the term in the 1980s.

So, a question for geologists: Does a new epoch make sense or not? The majority of the members thought, perhaps helped along by the wine and Zalasiewicz's friendly spirit, why not? The Anthropocene is at least an idea worth exploring. So they tapped Zalasiewicz to lead that investigation, a charge that has now spun out into the working group, a slew of scientific papers, and perhaps, ultimately, a life's work. "I think it's a good thing," Zalasiewicz says. "Hell, when has stratigraphy ever been popular?"

This fight is taken very seriously worldwide, partisans lobbing charges like "Marxist stratigraphy" at other partisans who think perhaps their interlocutors are merely sticks-in-the-mud. Stan Finney, one of the geologists ultimately in charge of determining new epochs, has laid out his case for what proof such an Anthropocene would require in a paper published by the Geological Society of London itself. He worries that the Anthropocene may be a political concept in search of ratification in the rocks. If it does exist, why not peg it to a date in human history rather than a particular record that may or may not appear in future rocks? This putative new epoch seems a projection into a future that we cannot know, rather than an already extant unit of the planet's more than 4-billion-year history. What good does this concept do people, anyway? Furthermore, what good does recognizing it in geology do? The concept came from the wider world and seems to get along just fine without formal stratigraphic recognition. When the first plants evolved to grow on land, there were dramatic changes to Earth's surface and atmosphere, yet "we don't have a name for that incredible revolution and period in Earth's history," Finney says. "We refer to it, the time during which it occurred, as from the late Devonian through the Pennsylvanian." Not the Phytozoic or Plantisian.

Geologists still lick the wounds from the fight over the Holocene, the long period between glaciers during which civilization developed over the last 10,000 years. If the Anthropocene is deemed old, it could perhaps supersede this designation—and reopen those wounds. There are also those who still think the Pleistocene reigns as humanity thrives in a long summer—made longer perhaps and definitely hotter by fossil fuel burning—before Earth's orbital peculiarities bring the ice marching back. No previous interglacial merited epochal designation.

Nor is the Anthropocene necessarily a new idea, just a new name for what was once derisively dubbed the Poubellian, the Plasticene, or the idea explored, a bit tongue-in-cheek, by the Upper Holocene Boundary Commission at the University of Iowa, among other outfits.

Finney is not alone in his skepticism. There are plenty of other geologic critics of this trendy term, including Philip Gibbard, chair of the Quaternary Subcommission and himself a member of the Anthropocene Working Group. That's part of Zalasiewicz's plan—he wants to include as many divergent voices as possible in this debate, ranging from skeptical geologists to ecologists and even journalists.

This motley crew seems to lean pro–people's epoch, as does the array of scientists who have adopted the term and used it to examine everything from biology to art. "It's a descriptive term for a set of phenomena,"

Zalasiewicz says. The reception from stratigraphers "hasn't been overwhelmingly hostile," he adds.

"Or overwhelmingly positive," I suggest.

"No, in the stratigraphical view, these things tend to be weighed up, chewed over for a long time," Zalasiewicz admits, not unlike the workings of the geology itself.

The proposals for how the Anthropocene starts in a geologic sense are numerous. There are the disturbed soils the world over that show farming and stretch back millennia, first noted by titans such as George Perkins Marsh in his magisterial *Man and Nature* (1864) and the Russian geochemist Vladimir Vernadsky, who called these human changes to the soils "a new phenomenon in geological history" a few decades later. The Holocene itself may have been prolonged by the invention of swampy rice farming in Asia, and the attendant production of methane, another greenhouse gas capable of warming the globe and holding back an ice age. Similar farming will show future geologists tree pollen replaced by grassier plants, like corn, the world over. Or perhaps our true mark will come in the extinction, like the graptolites, of most of the big mammals, starting 50,000 years or so ago. There are no more woolly mammoths, ground sloths, or giant kangaroos. The lethal pairing of hunting and burning by one species—*Homo sapiens*—meant the end for many megafauna.

But most proposals plump for something a little bit more recent, perhaps no further back than 1610, when mass deaths occasioned by the arrival of Europeans and their diseases devastated millions on the continents that would come to be known as the Americas. Thousands of hectares of former farms reverted to forest, the thinking goes, resulting in a drawdown of atmospheric CO_2 that culminated in the Little Ice Age back in Europe. Proponents call this the Orbis spike, though it is not as globally unified in time and space as some geologists would prefer.

Another marker might have an even more precise start: July 16, 1945. A radioactive isotope clock started ticking that day with the first test of an atom bomb. Rare isotopes—cesium 137, plutonium 239 and 240—created by that test and others like it will take millennia or more to fade away. There are no known natural sources for cesium 137; it would not exist in the world but for the manufacture and detonation of nuclear weapons. Those markers are still with us, and in us, like the strontium 90 in all our bones.

Or perhaps a better precise start date would be July 2, 1909, when the German chemist Fritz Haber first demonstrated his technique for wresting nitrogen from the air as ammonia and turning it into fertilizer (and the nitrates needed for conventional bombs). There would not be enough nitrogen for at least half of present-day humanity to exist without this fundamental chemistry breakthrough. Or there are out-of-this-world suggestions like the boot prints in Mare Tranquillitatis on the moon.

But the best marker for the Anthropocene, in the view of this non-geologist, is microscopic black balls of carbon scattered all around the world. Known as spheroidal carbonaceous particles to environmental scientists, as incomplete combustion products to engineers, and as soot to the rest of us, these billions of black balls can be found from the high Arctic to the bottoms of lakes in Chile. When a flame eats into a hydrocarbon such as coal or oil, not all the carbon atoms pair up with the oxygen that makes for fire and carbon dioxide emissions. Some of the carbon remains bonded tightly to its fellow carbons, unburned, forming these tiny black spheres.

This soot, fly ash, or whatever you care to call it is borne aloft by the hot gases of combustion before settling gently back to Earth and into lungs, lakes, the sea, or atop the land. The more coal and oil burned, the more of these tiny black balls, making such soot the perfect marker of humanity's growing pyromania. As the carbon that did pair with oxygen traps heat high in the sky, these pitted carbon balls mark the potentially irrevocable change. Similar though lighter soot particles can still be found from the volcanic eruptions that apparently aided the mass extinction at the end of the Permian period 252 million years ago.

The more recent dark balls appear almost simultaneously everywhere, around 1950 and the beginning of what some scientists call the Great Acceleration. That's when human impacts really took off, from the creation of chemical compounds unmade in the eons before humans to the population boom that put more than 7 billion people on the planet—exactly what geologists look for in determining the boundary of geologic ages. Our need for fire has left a consistent black smudge that covers the globe and may be remarkable even in the vast expanse of geologic time. Everywhere we step, we land on our own residue.

"If the word 'Anthropocene' wasn't there, I suspect we'd probably need a term of some sort," Zalasiewicz tells me as we sit in his office. "It's there. It's clearly not going to go away." And regardless of what the persnickety stratigraphers and geologists decide, the Anthropocene may be a story from the rocks that has already escaped their grasp. "I suspect the term, formal or informal, will persist," Zalasiewicz admits, perhaps a bit ruefully. Pluto is still a planet in the minds of many, and the Tertiary still exists as a period in some geology classes, even though the nitpickers of stratigraphy deformalized the term long ago. Which begs the question: How can you have a Quaternary, our present period, without a Tertiary before it?

Ultimately, the students will decide, as the old men and women of geology pass away. As the saying goes, science progresses one funeral at a time, and the geologists of today had to overturn the orthodoxy against plate tectonics among their own teachers just one generation back. The Anthropocene is attractive to students: "It shines a different light on geology," Zalasiewicz offers. "It shines a light or gives perspective on the world as we see it. You can help interpret the past, which is fairly abstract, especially if you're just getting into geology, by the example of the present and the way that something is fossilized or preserved or changed."

"So you can use it?" I ask. The concept does have utility to geologists?

"As a teaching device," Zalasiewicz says.

The hard record of humanity may not leave much. As Zalasiewicz writes in *The Earth After Us*, "One hundred million years from now, nothing will be left of our contemporary human empire at the Earth's surface . . . The Earth's surface will have been wiped clean of human traces" (unless we're busy laying down new ones).

Not much will be conserved, as climate shifts, species wax or wane, and the ever-changeable human nature changes priorities anew (or disappears into some deathless cyborg future). But river sediment, when not blocked by dams, will slowly bury technofossils—the delicate filigree of a plastic bag, the random metal deposits of a car, the formerly molten plutons and cracked rock surrounding an underground test site for nuclear bombs. Perhaps even whole cities like New Orleans, New York, or Miami will one day be buried in river mud and fossilized for future eons.

Of course, the focus of geology remains not what the ground can tell us of the Ordovician or some putative Anthropocene, but what use people can get out of the ground, particularly extracting the fossil fuels that represent the energy stores of billions and billions of years of photosynthesis: tiny dead algae floating down, escaping consumption, and getting buried in the river muck, or swampy forests and peat compressed into the brown or black burnable carbon known as coal. The planet bears a record of rot, decay, and mud slowly turned to something geologists can recognize.

The pursuit of such fossilized sunshine has reworked the landscape on a massive scale, more than a million oil and gas wells drilled in the United States alone, pinpricks in Earth's skin—and those are just the ones we know about. The special organizations of people sharing a common purpose known as companies have scraped away entire mountains in pursuit of coal, leaving a leveled and barren landscape in West Virginia and elsewhere, all in search of the residue of the Carboniferous—an entire geologic period of 60 million years named for its primary product, which we are busily putting back in the atmosphere. Zalasiewicz's students win prizes from oil majors for mastering subsurface-mapping techniques, like a detailed layout of southwest Wales showing folds in the rock like a cinnamon bun. Cheap, abundant energy in the form of fossil fuels made the Anthropocene—and modern civilization, from industrial farming to the human population explosion—possible.

All the off-gassing from burning will be recorded in the future rock, as will the slow pileup of silt and dirt known as sedimentation, inexorably leading to fossilization. Such records of the planet's past are laid out on Zalasiewicz's desk in a sprawl, like the various shales wrested from beneath the Spanish desert on his most recent student expedition, half hidden under a sombrero, complete with tassels and signatures in Sharpie. Geologists, like Zalasiewicz and his students, are storytellers, building tales out of rock.

Geology gives us the scientific creation story, complete with flood, fire, and extincting plague, all written in rock via graptolites and others—albeit a creation story that also needs cosmology to be complete. Some future alien visitor, as Zalasiewicz imagines, may note that we even got rid of all the other types of people on the planet: *Homo neanderthalis* and *Homo denisova* subsumed into our gene pool, according to the latest genetics. The fossilized skeletons the aliens find, if not the living humans, will show that *Homo sapiens* had swelled to 7 billion specimens by now, a rich fossil trove not unlike graptolites. If the corals go as a result of acidified oceans, that will be obvious too.

Or maybe the record starts when *Homo sapiens* began to transform the planet through agriculture—starting in the Middle East, China, and India, and slowly spreading from there. All those rice paddies and corralled cattle, sheep, and goats are perhaps enough to shift the climate anew, argues the paleoclimatologist William Ruddiman of the University of Virginia, or at least enough to shift the predominant future fossil pollen from tree to grass. At a minimum, most of the continents have been worked over for thousands of years, which is a history the land still holds. More than two millennia ago, Ashoka, the great emperor in India, introduced the idea of conservation—and it still hasn't stuck. In the end, agriculture took over the surface of Earth, just as John Stuart Mill predicted in *Principles of Political Economy*: "With nothing left to the spontaneous activity of nature; with every rood of land brought into cultivation, which is capable of growing food for human beings; [with] every flowery waste or natural pasture ploughed up, all quadrupeds or birds which are not domesticated for man's use exterminated as his rivals for food, [with] every hedgerow or superfluous tree rooted out, and scarcely a place left where a wild shrub or flower could grow without being eradicated as a weed in the name of improved agriculture."

As agriculture reached many of the limits of arable land, the pushback began, scarcity driving value, and national parks like Yellowstone were born, unique and rare—10 percent of Earth's surface—outposts and islands of conservation. Take, for example, the prairie of corn and wheat that covers the core of North America in place of the mixed prairie of grass and flowers that once supported thunderous herds of buffalo and antelope pursued by American lions and cheetah. Cows have taken their place, and the last big cat, the cougar, has slunk off to the mountains and woods.

An excess of pyrite (aka fool's gold) will accompany human traces as mobile iron and sulfur migrate to fill the shape of a former steel beam, plastic tube, or cell phone, a sheen in the rock that will show that we manipulated elements.

What you need from a fossil marker of time traditionally is a species that was widespread and then vanished. Perhaps humans will serve as that sentinel, a species that spread to cover the globe in record time and then dwindled or disappeared, like the graptolites before us. For some, that is the argument of the Anthropocene—a warning that our bad ways will quickly lead to our extinction. But for others, it's a challenge. How do we make a good human epoch—or even stretch it so it lasts for an era?

"Our actions will literally be raising mountain belts higher, or lowering them, or setting off volcanoes (or stifling them), or triggering new biological diversity (or suppressing it) for many millions of years to come," Zalasiewicz writes in *The Earth After Us*. "We have left our mark."

On September 17, 1939, the Russians invaded Poland. One of the first stops for the Soviet soldiers was the tiny town of Stolpce. In that town lived sixteen-year-old Irena Beresniewicz, along with her Polish father, her mother, originally from Smolensk across the Russian border, and younger sister Janina. Her older brother Olgierd, or Olek, was in the Polish army, serving in the north as Warsaw fell in a mere seven days to the German onslaught coming from the west.

In Stolpce, after several months of waiting and worry, the Russians evicted Irena's family in the depths of the winter of 1940 and loaded them onto trains headed east, liberating the countryside from the bourgeoisie like her father, manager of the local timberlands. The Soviet state carted them to Vozhega, northeast of Moscow, a work camp of "free settlers" exiled in the taiga in a town called into being just a decade earlier because of a new railway stop. The Soviet masters put the young girls to work cutting wood while the commandant tried to recruit Olek to the Communist cause. The bogs that lay in every direction prevented escape, swallowing those who tried, and the free settlers from Poland joined previously evacuated Ukrainian kulaks, the non-Russian locals, and other Bolshevik undesirables.

The Beresniewiczes and their fellows survived first one, then two brutal winters as another year of exile followed the first. Her father found Irena work on a nearby collective farm whose original kulak owners had been executed, along with their three children. She would count the workers, their time, amount of produce collected, and their pay, a shell game she struggled with under the angry eyes of her supervisor. But that did not last long, as the Germans invaded Russia that summer of 1941. As a result, the Soviet government freed the free settlers to go where they liked.

Olek set out first with the other young men to join a Polish army reorganizing for the fight against Germany in southeastern Ukraine. The rest of the family followed later, first going farther east by train, to Perm in the Ural Mountains, where Irena lost her family in the tumult. In an attempt to get to Orenburg, where she believed her family would be, she jumped another train to Sverdlovsk. On that train, she fell under the protection of a kind Russian officer, in charge of a traveling unit of the wounded, who bribed her way there. In Sverdlovsk, she miraculously reunited with her father, before traveling on to Buzuluk, where the new Polish army was gathering.

But this new Polish army was sent yet farther east and south by the Soviet authorities, to the ancient city of Tashkent in far-off Uzbekistan, and then farther still to the town of Guzar, where Irena herself joined the Polish army so she could care for her mother and sister after her father had to leave to join his own garrison.

In Guzar, her mother died of pneumonia, and they buried her in the desert, leaving the two young women, girls, really, Irena now all of eighteen, to fend for themselves. The Polish army began finally to move west, taking all of them with it. First Krasnovodsk on the shores of the Caspian Sea, then onto a steamer and to Bandar-e Pahlavi in Iran, a country that had only recently come under Allied control. By truck then to Tehran, where Irena signed papers sending Janina to an orphanage in the mountain town of Esrahan. Irena herself was graded by the military authorities and, because of a previously unknown weak heart, tagged Grade C, meaning fit only for office work and nearly thrown out of the army entirely.

As the Polish army continued moving west through the land of Persia toward Baghdad, she managed to find Olek again. He told her that their father had died of exhaustion, and perhaps a broken heart, far back on the journey. There was little time to mourn. Irena and the remaining Poles finally reached Palestine, and Jerusalem, where the 2nd Polish Corps of the British army waited.

The war carried on and, eventually, the Beresniewiczes made it to England. Olek became a physicist, helping build airplanes, and he later moved to the United States to work on the Boeing 747, then moving on to Lawrence Berkeley National Laboratory under the name Alexander

Blake. Janina survived the war in an orphanage in South Africa and eventually moved to California as well. Irena, meanwhile, met her future husband and future industrial chemist Feliks Zalasiewicz in Palestine during the war, married him on the shores of Lake Como in Italy a few years later, and had two children: Anna and Jan.

The family settled in Britain, in Lancashire. Jan Zalasiewicz owes his accent to that choice, and he is here to explore rocks and the concept of the Anthropocene only because of the family's epic trek. The residue lingers. He still speaks Polish, as I learned during a phone call in his office at Leicester University.

"Your Polish is good," I note, a bit surprised given the intervening decades.

"Well, it's charmingly broken, I think," Jan replies. "It gets me by."

"But you still speak it."

"I still speak it. My wife is Polish, so I had to kind of relearn."

"That's good. That's how you keep a language alive."

Zalasiewicz's son is now a young man, and he also speaks charmingly broken Polish, which he expects to improve as he spends time in Poland itself.

"Usually it's the third generation where it really . . ." I offer. As a grandson of an Italian immigrant, I speak no Italian. And it can be worse. All over the world languages wink out as the last native speakers die and the youth opt for today's largest lingos—Mandarin, English, Spanish, Hindi. While traditional knowledge is lost with those languages and words, the flip side, of course, is that more people can talk to each other.

"You need a few months in Italy, don't you?" Zalasiewicz offers.

Well, who doesn't? I think, fantasizing about Lake Como. Like most people, I would not be here without the luck of my forebears, desperate refugees one and all. Another hallmark of our species is motion, making our epoch the Kinesicene as much as anything, moving people, words, foods, customs, even plants and animals. Go far enough back and everyone is an immigrant, refugee, or invader. War is one of the fundamental human forces that reshape both civilization and the planet. War is also a force that has gone global, both destroying, as in the draining of the great marshes of southern Iraq in recent decades that may well have been the original Garden of Eden, and preserving, as in the demilitarized zone–turned–nature preserve that separates North and South Korea or the Green Belt that has replaced the Iron Curtain in Europe. I would not be writing this had my grandfather missed not one but two U.S. Navy destroyer escort assignments, ships that went on to be sunk in battle with Germany and Japan in World War II. Or my great-grandfather, rather than my great-uncle, could have been run out of Cleveland's Little Italy with his family.

We expend much of our creativity on concocting better ways to kill one another, whether warriors from the steppes mastering the skill of killing from horseback or today's intercontinental ballistic missiles tipped with nuclear death. The list of war inventions is long, including the Internet that helps connect the world today.

The only other font of innovation that approaches war is commerce, the first known writing a ledger of merchant's goods from Sumeria. In that sense, the human technology of writing as a means of storage started with the spreadsheet, a litany of grains, beer, and other cereal products. Ku Sim, this first note maker, knew barley and how to count and the symbols to record this accounting, without necessarily knowing math or even writing per se, even though this is the first preserved record of such notation, the direct ancestor more than 5,000 years ago of the words you are reading now.

The survival and eventual rediscovery of these clay tablets was random, just as survival in a global war is random, whether for my grandfather or Zalasiewicz's mother. Fate remains random despite the best efforts of those people who would control everything, just as there is still a certain randomness to which ideas become infectious, a meme, in the words of the militant biologist Richard Dawkins. Memes are ideas that propagate, and the Anthropocene is one, colonizing mind space in the early decades of the twenty-first century.

The term itself is not that old, popularized by Paul Crutzen, a Nobel Prize–winning atmospheric chemist. Crutzen found himself at a meeting of the arcanely named International Geosphere-Biosphere Programme in Mexico City in 2000, where paleoclimatologists discussed the environment of the Holocene, over and over and over. Fed up, the Dutchman stammered out in frustration: "It's not the Holocene. It's the . . . the . . . Anthropocene."

The idea spread like wildfire, helped along by scientific articles by Crutzen and Stoermer and eventually stretching beyond the sciences into the rest of academia, including the humanities. The meme penetrated popular culture through journalists like myself but also the songs of the independent bands Trans Am and Cattle Decapitation. There is Anthropocene fiction—Kim Stanley Robinson's Mars trilogy or Jeff Vander-Meer's Southern Reach trilogy—and Anthropocene film—*Snowpiercer* or *Mad Max: Fury Road*, among others.

Perhaps the Anthropocene is so popular because it's not new; our minds were, in a sense, prepared to accept it. The fundamental idea of an era shaped by people is an old, old one. "This seems to be a cyclical idea within geology," I note to Zalasiewicz.

"Oh, it's very much. Yes."

"It's been proposed before but never took . . ."

"It's never been taken seriously by geologists."

The idea has been hanging around in the Western world since at least the 1700s. Georges-Louis Leclerc (better known by the title his father purchased for him: the Comte de Buffon) was the first to divide the history of the planet into epochs in his 1778 work of geologic and popular speculation titled *Les Époques de la Nature*. It included a seventh and most recent one dubbed the epoch of humanity, or *Lorsque la puissance de l'homme a seconde celle de la Nature*, wherein humans helped shape Earth along with natural forces. Not quite as catchy as Anthropocene. But this slim supplement to his magisterial and uncompleted forty-four-volume *Histoire naturelle, générale et particuliere* was the first attempt at a science-based history of Earth, as opposed to a religious one, and the direct intellectual ancestor of Crutzen and Stoermer's modern meme.

Buffon also came up with the idea that species could go extinct, noting the many animals and plants that no longer existed that were preserved in the limestones around Paris. This was decades before his intellectual descendant in France, Baron Cuvier, suggested that extinction could occur in the Christian god's perfect creation, even though it is Cuvier who often gets the credit for making this intellectual break from then accepted theology. Buffon even used the same example—the mammoth, known by bones found in what would become Ohio—to show that some species had completely disappeared. He also knew that coal was the remnant of plants that died in long-ago swamps like those at the mouth of the Mississippi River, even if he was off by orders of magnitude in how long ago that was.

The Buffonian view is that human history and Earth history have become inextricably intertwined. It is a view that science has found its way back to, just as Charles Lyell, the father of modern geology, cited Buffon in his *Principles of Geology.* Nonetheless, as Lyell reports, his contemporaries found these speculations "reprehensible, and contrary to the creed of the church."

The idea has many more parents, of course. The Berlin aristocrat and naturalist Alexander von Humboldt called life as we know it the *Weltorganismus,* world organism. Antonio Stoppani—geologist, lover, priest—called the present age the Anthropozoic in his *Corso di Geologica,* noting again the recent disappearance of all those big animals as well as the signs of crude agriculture stretching back at least 10,000 years. "The creation of man constitutes the introduction into nature of a new element with a strength by no means known to ancient worlds," he wrote. "This creature, absolutely new in itself, is, to the physical world, a new element, a new telluric force that for its strength and universality does not pale in the face of the greatest forces of the globe."

The polymath George Perkins Marsh, the oft-forgotten father of American conservation and a member of the Board of Regents of the Smithsonian Institution, includes the Anthropozoic in *Man and Nature; or, Physical Geography as Modified by Human Action* while his fellow geologists still argued over exactly how old the planet was. And in 1854, the British geologist and theologian Thomas Jenkyn described the present day as "the human epoch" in published lectures. Steam power, the railroad, and other modern powers convinced these citizen-scientists of the power of man, and it was always man in those days.

The professor of geology and natural history Joseph LeConte at the newly founded University of California at Berkeley preferred to call this newest period the Psychozoic because it was really the life of the mind that was novel in this new reign of humanity, a term also employed by the great Polish futurist Stanisław Lem in his writings much later.

Still, the international Third Geological Congress in 1885 definitively chose Holocene to replace Lyell's "Recent." Whatever the name, it always sounds better in Greco-inspired English. Then the Holocene slowly but surely, like mountains ground down to sand by the rain, came to predominate in Western geology, though it took until 1967 for the U.S. Geological Survey to formally recognize the term and tie it to the modern migration and flourishing of humanity.

It was the discovery of radioactivity in the twentieth century that allowed scientists to begin to estimate the true age of Earth with some precision, adding strength to Lyell's theory of ageless ages where rivers flowed and carried sediment to the sea, where the sediment piled up and turned to rock over eons only to be thrust back to the surface by the mysterious workings of shifting plates, then weathered by rain and carried back to the sea by rivers to start the process anew. The radiochemist Bertram Boltwood used radioactive decay to measure the age of rocks and minerals to come up with an estimate of 2.2 billion years old in 1907, an estimate later doubled by more precise techniques. By 1913, the British geologist Sir Arthur Holmes published the first quantitative geologic timescale that, over revisions through the 1940s, stretched back more than 4 billion years—and ended in the Holocene.

But the Russians kept this idea of human influence alive. The geologist A. P. Pavlov's "Anthropogene," covering the last 2.5 million years, was a replacement for what Western geologists call the Quaternary, which is simply the fourth period of this Cenozoic, or "new life," era. Cyclical ice ages define the Cenozoic, a naming that was still debated as late as the 1980s at international geological congresses.

But the idea of humanity's influence on global nature is even older than all this modern science, stretching back at least to the religious belief in one all-powerful god. In monotheistic religions, humanity is placed at the center of the world, even the center of the universe. We often judge ourselves to be the most important and most blessed species. So the Anthropocene epoch of Crutzen and Stoermer is just another episode in the eternal recurrence of an idea born out of anthropocentrism.

"I remember the joke lectures about the Coca-Cola bottle era and all of that," Zalasiewicz says to me, and I imagine him in the field with his jacket bearing leather elbow patches, a piece of straw firmly clenched between his teeth, musing on a rock in his grasp. "These were fringe ideas. It wasn't serious. It wasn't real."

But now it's showing up in the rocks, like the plastiglomerate formed from plastic residue, lava, and sand on the beaches of Hawaii.

"It is kind of a messy idea for geology," I offer.

"It is. It is messy. It's unfinished. It is ongoing. It involves all kinds of criteria which are not normal geological criteria."

If there's one thing geologists hate it's ideas that are not buried in deep time. "How would this be a useful concept and not a nuisance for geologists?" I ask.

"It's in use and therefore it's not an abstract or artificial concept. It's not trying to plug in, it's not trying to radicalize the whole science. It's just simply there."

"So the Anthropocene looks different from any other geology?" I ask.

"Yes," Zalasiewicz avers, noting the complexity of the shapes in the rock of what humans have left behind and will leave behind, its patchwork nature and shapes embedded in more typical layers of rock. Our churning of the subsurface, known as anthroturbation, like fracked wells and gold mines, distinctive in the last 4.5 billion years. Our burrows—mines, boreholes, tunnels—put the earliest bioturbations of the first multicelled creatures to shame. We are the first animal to make a global mark. "I think there's a sort of novelty in there, and that is part of its defining features, of course. It also makes a comparison because, if you have no close analogs in the past, then you're looking for distant analogs and sometimes clutching at those straws, once one has finished chewing them . . ." Zalasiewicz trails off.

"I don't know, straw seems to be in short supply judging by the train ride up. There's not a lot of it around. The sheep are eating it and . . ."

"It's the bloody Anthropocene," Zalasiewicz concludes. Indeed, especially bloody since some tie the start of the epoch to the mass death, slavery, and warfare that wiped out many of the humans on three continents.

Massive technofossils like London and Shanghai will call out to the future: Something was here! "That's an awful lot of stuff that's piled up," Zalasiewicz tells me on that cloudy fall day in his office. "Let's say the sea comes in and that's preserved."

"Seems to be coming," I note, thinking of the meltdown of the ice sheets in Greenland and Antarctica swelling the seas.

"It seems to be coming, yes. So that will be tens of meters thick; you can't squeeze that to nothing." Mud compresses under geologic pressure by as much as 90 percent, but sand is far more resistant, squeezing out water to compress by maybe 40 percent. "You'll have a thick band in some parts of the world, where, say, New York was," he continues.

There will be all the stuff we've removed: 3 billion metric tons of coal

On the same day that U.S. president Obama discussed new allegations of the federal government spying on Americans, a small band of scientists discussed a new paradigm for "earth system governance" and how to stop undesirable actions such as Chileans burning Patagonia. The Anthropocene is truly an age of fire, notes Pauline Dube, a wildfire specialist at the University of Botswana: "Thanks to humans, the whole world is now at wildfire risk," even those places that did not burn before, like Israel or Russia. Fire is everywhere in the world, some of it hidden inside machines small and large, some of it visible from space, vast smoke plumes fouling the atmosphere and raining ash over long distances, the red infrared glow of the blazes themselves.

Hailing from the world over, geologists, physicists, and oceanographers debated with sociologists, anthropologists, and philosophers the very definition of "Anthropocene" as well as what the word is meant to achieve or describe. A soft-spoken wizard with long hair, the hydrologist James Syvitski, studier of rivers as reshaped by people, presides with a constant grin, even when debates turn personal. His wizardry includes movies showing deltas enlarging as forests are cleared and sediment washes downriver, streams whose colors are changing as gold is mined, land sinking as sediment no longer flows down the Mississippi and the Yellow Rivers.

Some argue that the Anthropocene is a "doomsayer," the herald of the end, while others see the opportunity for solutions. The Anthropocene at least offers the chance to reconcile physical and social science, bringing together the puzzling out and tinkering with the nitrogen cycle as well as the social and economic relationships that govern its use today. The key to an enduring Anthropocene will be to preserve as many options as possible and to use the term as a tool to prevent decisions from being made with a singular focus—biofuels for energy security that end up worsening food insecurity and environmental problems, dead zones, erosion of precious soil, and the like.

Hovering like a specter in the background of this meeting is the long-ago work of "limits to growth," planetary boundaries that have yet to be fully discovered but that once crossed mean precipitous, perhaps catastrophic, change. Skeptics, including the Australian ethicist Clive Hamilton, whom I'm sitting next to, sit back, breathe deeply, and ponder whether the very idea of the horrible truth of the Anthropocene changes anything that matters, like politics, or whether there is any hope of management of this dangerous place and period for all life on the planet where, for the first time in Earth's history, geology is being influenced by a conscious being, marking not just a geologic shift but also a social change. A phrase that might become simply a tool for denial of this urgent existential threat, or as Hamilton says, "If knowledge is power, it is also the power to do nothing."

The Anthropocene requires the development of a new human perspective, a kind of fourth person who views the world from the outside and notes what "we" are doing while at the same time recognizing that the "we" is "you." "You are screwing yourself," Karen Seto, a scholar of urbanization at Yale, notes.

At the very least, the Anthropocene has outpaced the ability of scientists in their separate disciplines to keep up and must avoid the trap of suggesting that all previous generations were somehow unsophisticated and incapable of noticing the change or that science alone is the solution. Perhaps even science itself must change from being mostly question driven to more solution directed, and those solutions will most likely never be global, like the fever dream of geoengineering. Just as the problems emerge locally but aggregate and swell to impact the globe, the real solutions will have to be adapted to local circumstances and scale up from there.

The goal, if there is one, is to minimize the negative, just as there is no such thing as a sustainable city, just more sustainable practices for making and running cities. The solutions on offer here so far are insufficient: a tax, whole Earth accounting, some kind of label, Earth governed like a condominium, and humans somehow organizing the global neighborhood. The one belief that makes the Anthropocene matter is that life, all life, is precious.

Arguments rage online and in the real world about a good Anthropocene versus a less bad one, or whether the Anthropocene even exists and, if so, who should get credit and blame for it. It is bickering among a species that has exploded into prominence in a geologic blink and may disappear just as quickly.

The road to hell is paved with good intentions, it is said, like the campaign to treat diseases with injectable medicines using unsterilized needles in Central Africa that helped the evolution (and spread) of HIV. Or the endless pursuit of oil or natural gas that leads to earthquakes. If there is any doubt that humanity is a geologic force, the fact that our thirst for oil can literally shake the ground should dispel it.

Zalasiewicz sums it up: "You're not going to understand the Anthropocene without understanding people."

The disappearance of species is common in the vastness of geologic time. An estimated 99 percent of all species that ever lived have died out. But the pace of extinction seems to be unusually fast now; by some estimates, animals and plants are dying out 1,000 times faster than the ebb of species in the past. The fossil record suggests that in a normal year, somewhere between 1 and 10 plants or animals went extinct for the roughly 9 million species (that we know about) on the planet. Now thousands do. Even so, life is hardy. Rats famously survived forty-three nuclear blasts on the test atoll Eniwetok in the Marshall Islands.

But it is extinctions in the fossil record that stratigraphers most often use to date and name specific eras. Instead of absence, it may be presence for the Anthropocene, and human skeletons are not the only candidate for index fossils. For Zalasiewicz, that means his cat, Mimi. For every one of the 3,000 or so tigers now alive, there are roughly 100,000 house cats, eating tuna from a can at human hand. As the big cats disappear, the house cats thrive, and that is the kind of population explosion that is captured time and time again in the geological record. Perhaps the skeletal marker of the Anthropocene won't be human remains in abundance but those of our pets or other domesticates. Cows, pigs, sheep, and people now make up 95 percent of the weight of all animals on Earth, bulked up even more thanks to the extra weight afforded by human tricks like pulling nitrogen fertilizer from the air.

Our preference for one animal—the cat—means doom for other animals—birds. Cats, both domestic and feral, kill more birds than does anything else, including human hunters, human-built wind turbines or, the next biggest killer, human-made buildings sheathed in glass. Cats, like humans, are on every continent now, including Antarctica, and there may be as many as 500 million *Felis catus* extant today, maybe more, since no one can take a reliable census. A future full of cats is perhaps to Mimi's liking, less so for the songbirds and ailurophobes—and cats may outlast even their current masters, surviving perhaps a human-induced mass extinction of humans and, if loosed from human constraint, freed to evolve in resplendent forms in any post-Anthropocene. Human choices have consequences in directed evolution. Some extinctions are even appreciated, like smallpox, or possibly the saber-toothed cats that preyed on our ancestors.

This modern epoch is trending toward the homogenous, a particular mix of cats, rats and mice, pigeons and seagulls, as well as roaches and lice— creatures that travel with people wherever we go. Plants have been reshuffled, with tropical species growing in New York City and South Pacific pines and other plants gracing islands in the equatorial Atlantic. This process really kicked off in 1492, when the Old and New Worlds, Eurasia and the Americas, collided, prompting an interchange of plants and animals. From that Columbian Exchange, named after the great and deluded Anthropocenist Christopher Columbus by historian Alfred Crosby, the potato has spread across the planet, to name just one world-conquering plant species.

Then there's the fact that we might need to move even more species around as a result of climate change—or simply to preserve certain functions nature provides, like clean water or timber. Such managed relocation, assisted migration, or whatever other jargon scientists see fit to come up with makes eminent sense to some—and flies in the face of stewardship for others. Entire people's ways of life and habitat may perish too; just think of the Inuit in the rapidly changing Arctic.

The cost of all this has been wilderness, in the sense of pristine places untouched by the guiding hand of people, and, even more so, wildness, in the sense of plants, animals, and entire jungles, which has been pursued and subdued, chased in every form out of every corner of modern life. For good reason: tigers still eat villagers in India and leopards hunt their dogs. But some of those who lost wildness long ago—Americans, Chinese, Europeans, and Japanese—long for it. Modern humans suffer from "ecological boredom" brought on by our heedless pursuit of the "providential principle," pursuing natural resources, such as the cod off Newfoundland, as if it were endless until it is gone, in the words of the eco-thinker George Monbiot. At the same time, it's important to consider who pays to have this wildlife around, most often at a comfortable remove, and who pays by suffering the depredations of animals roaming free, as well as those who will pay to kill them.

the atomic bomb, where scientists are meddling with the world and doing things that are beyond the control of humans?" Of course, that's also Williams's proposed start date for the Anthropocene, all those isotopes marking the seeds of distrust between humanity and human scientists as well as the moment when human science can be detected for all time.

Even if CO_2 emissions stopped tomorrow, it would take perhaps 100,000 years for the billions of metric tons liberated by fossil fuel burning to disappear from the atmosphere. The plants and animals extant on the face of Earth and in the depths of the seas will have been permanently changed, disappeared, or redistributed, setting evolution itself on a new course for perhaps millions of years, which will also be patiently recorded in the rocks. Rocks have been picked up and moved from place to place, continent to continent, in a movement bigger even than the world-girdling ice of the Ordovician. Humanity is an unprecedented, erratic force in the geological sense.

The smoke signals of these present times will last for a long time too. Long after other worlds have been settled or civilization wiped away, a thin black line will mark one species' world-changing ways, pretty much everywhere. Cleanup efforts may make that line thinner or thicker, but a dusting of black will lie there, hinting at the cause of a climate change that will also be patiently recorded. Readers of this book don't need to be welcomed to the Anthropocene; we've been living in it for most or all of our lives.

"We are spoiled for choice," Zalasiewicz says of the effort to define a start for this new epoch. "Geologists simply try to do the best they can to define a boundary that works reasonably well reasonably often."

"It doesn't mean that's the big event," Williams adds. "The big event might still be yet to happen."

Geology might also offer an answer for how to turn the Anthropocene into an Anthropozoic, something a little more lasting, like the so-called Boring Billion, which shows how life can keep the world stuck or balanced. This vast span of time showed little change in the mix of life, rocks, and chemistry that drove the planet, though it paved the way for the rise of the animals. It would behoove humanity to try to secure that much space in time, but it may not be in human nature.

People are drawn to the dramatic. But there's actually something rather dramatic, at least from the perspective of that attenuated buzzword "sustainability," about the world's most boring geology. Of course, there's an even more boring prospect, the end of all life on the planet, an unlikely outcome from our perspective but not an impossible one.

Technology has grown to the point of posing existential risks: nuclear weapons, rejiggered biology gone wrong, artificial intelligence that treats us as we treat animals, the insidious threat of climate change as a result of fossil fuel burning. Implicit in the idea of the Anthropocene is an end, either because the Anthropocene lasts for the rest of time, morphing into an Anthropozoic or the like. Or because humanity does not last and there is no one to see the record in the rocks.

The Anthropocene is an attempt by the makers of the geologic time-scale to write our species into it. The record in the rock will wait, moving deeper underground, for those with eyes to see it, but it is up on top, on the surface, where that record is being written. And the end—whether a species abandoning its original home for other planets, a catastrophic space rock collision or plague, or the end of Earth itself when the sun swells to swallow it—awaits us all. That's a lot of weight for the story in the rocks to bear.

Part II

Terra Incognita

Chapter 3

Ground Work

There's not one atom of yon earth
But once was living man;
Not the minutest drop of rain,
That hangeth in its thinnest cloud,
But flowed in human veins . . .
Thou canst not find one spot
Whereon no city stood.
—PERCY BYSSHE SHELLEY, QUEEN MAB

Drones crowd the skies, confusing everything from birds to airplane pilots. I confirmed this on a balmy summer day from my perch on a small rise overlooking a children's day camp. While the little kids played soccer, I watched college kids run through a preflight checklist with the pseudomilitary rigor of remote-control-vehicle enthusiasts. "The only part I have to do to get it to fly is I have to enable the copter, I have to throttle it up just until the propellers spin and then I just flip the switch and it goes into auto and takes off," explains the mechanical engineer in training, Stephen Gienow, who dreams of working in the aerospace industry on drones or, as he calls them, "unmanned aircrafts." Unmanned is the theory, if not quite yet the practice.

Many of the little kids watched the drone on the ground more than the ball. Drones entrance people, even a hexacopter with the bottom half of a Tupperware container to shield its computerized brain and bear its name on a piece of masking tape. The *Crimson-I* can barely loft the Canon camera clipped to its underbelly as well as its own battery, which permits flights of twenty minutes or so.

Still, the *Crimson-I* has done twenty-five flights in a row without crashing—a new record. "I haven't seen the full auto yet," explains the ecologist Erle Ellis, mastermind of a scheme to use drones to monitor forests. If any of the novel woodlands, fields, and wetlands out there end up conserved, the first thing required will be keeping a better eye on them, and drones may prove a cheap, accurate, and easy way to do so. In fact, they're already being used to monitor everything from forests to the size of seabird colonies, not just killing people.

Most crashes occur right at liftoff, but that's when there is a human pilot. Ellis explains, in words that could also stand in for the state of conservation in a world still losing much of its old-growth trees, tundra, and territory, "You have to understand, it's nerve-racking this stage, usually." It's a bottleneck. The thinking with drones is: Remove the human pilot, reduce the chance of human error.

The light is bright and strong on the hilly campus of the University of Maryland, Baltimore County. Most of the trees simply line the walkways, so there's not too much shade to block the sun. This commuter campus opened in 1966 and remains under construction. More trees will come with time, no doubt, but in the meantime there is the Knoll, an old wood-lot of the farmer who sold his land to the university to build this campus. The Knoll registers as a mound of leafy green in the distance behind where the new pool building rises.

Fortunately, this summer day is free from the oppressive heat and humidity common to the Chesapeake region under the old Holocene or "long summer" climate regime, slated to get hotter and wetter in the new people's epoch. Because I'm going to have to go into the Knoll today, and not for what seems to be the most likely reason people go there: to find a secluded space to make out or smoke pot.

The whirligig, following the coded instructions of its Arduino processor brain, whispers up to speed, emitting more and more of a buzz. The Knoll is less than a mile away, but to the hexacopter and its team, the distance is much farther. That's because these little test flights are intended to prove that the drone can take the pictures needed to help save this forest, or any forest.

I am here in Baltimore to observe Ellis's team, which, not unlike the hexacopter itself, is cobbled together from spare parts, a mix of grads and undergrads from the environmental sciences and computer sciences.

All bent on proving that this concept can be used to monitor a wild forest and enable better conservation, even if the bearded, bespectacled guru of this ecology by drone is perhaps most famous for the concept of anthromes: the idea that the land and the interlocking mesh of life in it and on it—microbes to ants to fungi to plants to birds to squirrels to coyotes to us—are now shaped mostly by human influence. Anthromes range from "urban" to "wild woodlands," and they occur on all the continents, except in the minds of those ecologists and conservationists who think the word is Orwellian doublespeak. "Anthrome" encompasses the idea that the wild, in the sense of land free from human influence or direction, is now limited to the most remote and hostile spots.

The Knoll is not exactly a forest, but the former woodlot approximates a forest. The Knoll boasts thirty tree species. The local charismatic megafauna is mostly squirrels and maybe a few deer (or, more likely, college kids, depending on what counts as "fauna"). But the wild Knoll also takes a toll on the drones. Before the new auto takeoff and landing, Gienow or whoever had to pilot the delicate craft from a clearing amid the trees. Many a drone was lost to an ill-timed breeze, an errant move by a pilot, or a grasping branch.

In the old days, it was even worse. Ellis and his crew started off flying kites.

Mammoths likely once trod the land now known as the Knoll, which was much farther inland then, long ago in the icier clime of the Pleistocene. There were also dire wolves, saber-toothed cats and lions, giant beavers, and maybe even giant ground sloths. By the time the ice retreated far, far away, along with perhaps the mammoths and other big animals, humans had arrived. Evidence of the earliest inhabitants has been found in what are now Pennsylvania and Virginia, appearing perhaps as early as 16,000 years ago but certainly by 10,000 years before today. These humans reshaped the landscape of eastern North America, creating a broad oak-dominated woodland. And if humans contributed to the disappearance of those mammoths, tipping a vulnerable prey species over the edge by introducing a new, rapacious, and clever predator (us!), well, then, perhaps the Anthropocene is actually thousands of years old, a view supported by some paleoecologists.

Perhaps those sloths had it coming. By the time Europeans arrived, what would become Maryland was a kind of paradise for people. Fertile soils had help from abundant fish used as fertilizer. Sprawling woodlands supported the critters whose fur got the colony of Maryland started, economically speaking. English settlers came in *The Ark* and *Dove* but bore more disease than peace or animals. By 1675, the Piscataway tribe of 2,000 dwindled to just 150 people. The newcomers lived in fear of the deep, dark, seemingly endless woods. I'm sure they would fear the superhighways of their descendants even more—an unrecognizable change in a scant two hundred years.

Tobacco invaded with the Europeans, pushing aside the fur trade as the once mighty forest also dwindled as land was cleared. Agriculture has always been humanity's most widespread method for despoiling the world. Since at least that time, essentially continuously, the Knoll has been what some call a working landscape, shaped to suit the preferences of its human inhabitants. This is what Ellis, drone surveyor and consorter with archaeologists, calls an anthrome.

The neologism takes its root from an older neologism: biome, also known as an ecosystem—a grouping of animals, plants, and other life sharing a similar climate and geography, such as rain forest, grassland, or desert. Perhaps the most famous biome is the Mediterranean, the "center of the world" in the original Latin, which can be found in California, southern Africa, and parts of Australia, as well as the region that lends the biome its name. The plant ecologist Frederic Clements coined "biome" to describe these kind of superstructures of organisms that spread across a particular landscape.

In the case of the Knoll, the biome is the temperate broadleaf forest. That's the kind of forest that once stretched from the East Coast of North America all the way to the Mississippi River and one that doesn't exist in exactly the same way anywhere anymore, another recent extinction at the hands of people.

F. Scott Fitzgerald tried to imagine that lost history of the temperate broadleaf forest in *The Great Gatsby*: "Its vanished trees, the trees that had made way for Gatsby's house, had once pandered in whispers to the last and greatest of all human dreams; for a transitory enchanted moment man must have held his breath in the presence of this continent, compelled into an aesthetic contemplation he neither understood nor desired, face to face for the last time in history with something commensurate to his capacity for wonder."

That may or may not be true, but to get some sense of how the temperate broadleaf forest worked in the past, just head up the road to the

Smithsonian Environmental Research Center (SERC) in Edgewater, a relatively intact bit of woodland where scientists study everything from the exchange of energy between the tops of trees and the air above to the latest in portable laser measurement technology, which looks a bit like wandering the woods with a small picnic table attached to your waist. Still, the complete functioning of an intact biome or even an anthrome remains almost as unobservable as the secrets of quantum mechanics. The very act of observation, i.e., human intrusion, changes the thing observed. But, to simplify, from microbes deep in the soil to the upper-most canopy of treetops, all life works together and creates a rough equilibrium continually racked by boom and bust. At least that's the theory.

As human use became so much more intense in the region, pockets of conservation remained. SERC is a more consciously conserved land. The *Crimson-I* has flown here too, capturing in the full three dimensions the architecture of trees, undergrowth, and the contours of the land. The question going forward remains what to conserve, what merits saving, just as it was when aristocrats set up the hunting preserves that kept the Polish woods of Białowieża as the last remaining example of primeval European forest, albeit lacking the characteristic megafauna of aurochs and wolves, though it still had the forest bison for which it was originally created as a home. The iconic national park Yellowstone is set up as a land before people—for people to visit. And the Arctic National Wildlife Refuge—that totem struggled over as a last refuge and a last reservoir—won its conservation from the U.S. government through arguments for the global good.

The real biome of Maryland and much of the temperate broadleaf forest in North America these days is the city and the suburb, paved roads and the lots in between. The main plant is grass. The suburban lawn is a human-directed biome, a result of the choices made by one species—and a rival to corn as Americans' favorite crop. Hence Ellis's coinage: *anthro* from the Greek for "man" plus *biome* equals anthrome.

Even the Knoll qualifies. Prior to its incorporation in the campus, its role was as a kind of living fuel tank, the answer to what eighteenth-century Europeans bemoaned as the wood crisis, not unlike worries about peak oil in this latter age. The oldest kind of conservation is this kind of forest conservation, fuel reserved for later use. Side benefits include fresh air, greenhouse gases extracted from the sky, a home for squirrels and other animals, richer soils, even cleaner water.

The Knoll is not the kind of place anyone would particularly choose to conserve. A road cuts through it and the edges have been thinned. It's more of an afterthought in the landscaping of the college, a bit of tall green relief from the swards of grass and the squat brick buildings. But it still offers crannies and nooks to explore, fallen logs and their fungal kingdoms of decay, air redolent with the tang of leaves, even the sense of being in a natural cathedral of sorts as the trees reach to the sky and filter the sunlight. It is no redwood grove or Adirondack thicket, but it's about as wild as things get on the oft-used lands of the East Coast.

Each generation of Marylanders thinks of what it inherits as natural, whereas the landscape has been shaped—most often degraded, sometimes enhanced—by all the generations prior, back to the first humans to trod the continent now called North America. There is no pristine land. Biomes are always changing, if not as a result of climate change, then because some new animal or plant has stumbled onto the scene and changed the scenery. But also: Things can always get worse. That's certainly top of mind as Ellis and I sit together on his wraparound porch on another unusually temperate summer day, blessed by a freshening breeze and with only the singing of the cicadas and the occasional ripple of a wind chime to interrupt our conversation. "There is a pristine myth, but pristine is not a myth. It has existed, you know," Ellis tells me as he sits back on the wicker couch and sips iced tea. A butterfly flutters down and alights on the table between us, as if a staged demonstration of the pristine in the profane. "From our perspective, the pristine never really did exist," he adds, because wherever people go, change follows.

The resilience of nature to change is profound. The entire East Coast of the United States, once logged or cleared for farming, is a verdant forest again. The deer prove it, as does the sequestered carbon dioxide. The Knoll is different from what the first Europeans found. But it is very much alive.

"We are really good at extincting mammals, some amphibians. There are some taxas that tend to be more sensitive to human changes," Ellis goes on, his surfer dude speech patterns more reminiscent of his Santa Barbara grad years than his roots here in Maryland. "But the main effect that you see—and this is the reason why novel ecosystems are such a big deal and so important to this concept—is you just end up adding a lot. You are not taking everything away. You are shifting everything around completely." There is plenty of evidence to support this; for example, people's house cats have usurped the role of foxes in the modern Maryland suburban anthrome. Though maybe the foxes have a shot at coming back too.

"Hi, I'm Dana Boswell. Can I work for free?" That's the kind of greeting in the grim cinder-block-and-fluorescents ambience of the halls in the Global Information System department at UMBC that gets you sent into the woods with sticks and tape and nails. The goal is to inventory trees. More particularly, to inventory 7,300 trees at chest height to ensure that a drone can do as good a job, or better, without the need for such grueling work.

It is indeed grueling. During the summer, it was short but strong Boswell, with her sunny disposition, ready smile, and iron will, plus two dudes trying to figure out the best method to lay tape in a straight line in a forest full of undergrowth, brambles, and other impediments. The job started at six every morning, listening to the birds call out in those same trees before the university groundskeepers filled the air with the sweet smell of fresh-cut grass and the din of mowers. A bag lunch from home was consumed within the Knoll, scratching at poison ivy bumps and bickering over technique. Bugs swarmed over sweaty bodies, and temperatures climbed as high as 38°C. Even the dead trees had to be included.

Every day was the same, even in the rain: eight to ten hours, five days a week, then home to rest, put calamine lotion on bug bites and other itchy bits, and get up early the next morning to drive in and do it all over again. By the end of the summer, the ragged little team had finished surveying maybe 75 percent of the Knoll and half of the other test plot, known as Herbert's Run for its creek. Most of the black gums, tulips, poplars, oaks, beeches, hickories, and both green and white ash duly noted.

So that fall, winter, and into spring, and the first few weeks of the following summer, a group of three women led by Boswell tried to complete the survey, trudging out even in snow in the spare bits of time afforded to commuting college kids. Plus, the National Science Foundation funding stipulated that they could work no more than twenty hours a week. It takes thirty minutes just to lay out the tape in a forest and that's before the surveying even begins, so progress was slow, very slow. The "stupid-ass cold" didn't help, nor did the perhaps questionable dedication of people working for free. "It's really hard to fire a volunteer," Boswell notes.

On the plus side, the leaves were off the trees, allowing accurate measurements of the height of every tree with a kind of laser pointer measuring device known as the hypsometer.

Summer came again and, with it, the need to go out and redo most of the work, perhaps because of the questionable dedication of volunteers (though volunteer Boswell shows no signs of questionable dedication herself). She even had to enlist the aid of those more comfortable tinkering with robots or drones to come out into the forest and at least hold tape or count off a quadrant. Finally, one semisenior undergraduate who had taken at least one course in botany was designated the official species identifier, whose job was to go out into the forest and decide what was and was not a tulip poplar or swamp oak.

Only someone who once wanted to be a park ranger so she could shoot people who litter could have delivered the Knoll so fully surveyed. Boswell was that person, someone who "wanted to inflict serious injury on people that would disrespect Earth." Her childhood vacations were not to Disney World or the local mall, but national parks, including Big Bend and Arches. The woods or the desert were the only places where her perhaps hyperactive mind could quiet. (That hyperactive mind—and a youthful obsession with the Grateful Dead—is probably what explains her peripatetic college career, with stints in Washington, California, and Florida before ending up at UMBC at thirty years of age.)

"It doesn't seem like it'd be that hard to put nails in trees and measure them, but it's hard to do in a forest," explains Boswell. That's why the hope is for drones to be able to obviate or at least minimize the need for this kind of work, while allowing the same kind of measurement and thus management, or at least monitoring. In the meantime, the volunteer survey brigade got some "Ecosynth" coffee warmers as a thank-you parting gift, the green logo on the side a series of flat pictures capturing the details of a fully fledged tree with nary a human in sight. Or a drone, for that matter (making the logo a bit difficult to parse).

Meanwhile, Boswell has learned to fly the drone—and swears she'll never do another field survey, though she may order future minions to do so. She's gotten good enough, or the software has gotten good enough, or both, that she recently flew twenty-five surveys without a single crash.

Her time with Ellis has convinced her that her dream born of the national parks is dead, an anthrome among anthromes designated as natural despite being an artifact of some particular human desires. "That idea that I had of saving the planet from humans, that's kind of gone," Boswell says. "But nothing's really replaced it." And that's the present philosophical crisis of conservation in the proverbial nutshell, to go along with the real-world one that sees the end of biomes and even anthromes.

It's not easy to fly a drone, but it's even harder to use a kite.

Kites are how Ellis and the geospatial scientist Jonathan Dandois got started with such surveys. Dandois had been working with Ellis ever since he started classifying land types from satellite pictures as an undergraduate more than a decade ago. He then accompanied Ellis to China for ten weeks, zipping around the countryside, mapping and identifying the "anthropogenic world," the planet as made over by humans. "I guess I really liked that view of the world," Dandois says now, somewhat sadly, though that may just be post-dengue tiredness. "I have to appreciate this is the world we live in. It's not all wild and it's not such a bad thing."

That hard undergraduate work infected Dandois with the academic bug, but not so much that he didn't take a few years off to muck about, hiking the Appalachian Trail, assisting architects, and working on a ranch out in New Mexico, before slinking back to UMBC to complete his training as a mapping specialist. "I came back because I didn't know what else better to do with my life," he says, chuckling. "That's why people go to grad school."

Once safely ensconced back in academia, Dandois became the lead of Ellis's crazy dream: to use software to knit together these various maps to identify anthromes. The software would use a multiplicity of pictures to render a three-dimensional map of the terrain, a kind of artistic lie overlaid with human boundaries and human needs that nevertheless enables people to make choices, whether to save a copse or understand how paved roads lead to water pollution.

The problem was getting all those pictures. Dandois and Ellis's first idea was to use a styrofoam airplane. "We spent one summer just flying airplanes and crashing all the time," Dandois recalls. So he set off in pursuit of experts, ranging from a chemical engineering student at UMBC who flew such planes as a hobby to folks at NASA, expert in what's called remote sensing. The NASA experts warned him it would take at least a few years to become skilled enough in flying said planes to even attempt a picture, let alone the thousands of pictures such an effort would require. So they suggested, why don't you go fly a kite?

Unfortunately, "it's hard to use a kite for aerial photography," Dandois tells me, in his little office tucked away at the back of another basement cinder-block cell. "It's not repeatable, it's very hard to get the exact same thing."

Even worse, it's hard to judge where you are over the forest when you are far enough back to fly the kite, a thousand feet of line or more, and down on the ground. And even worse than that, from the perspective of the ultimate goal of surveying the woods, is that "you can't use the kite in the closed-canopy forest."

What Dandois needed was autonomy—a robot that could execute a flight pattern precisely tied to the Global Positioning System offered by the satellites far overhead, snapping pictures all the while. Fortunately, the relentless refinement of technology has delivered such a device: the hexacopter drone. So the original kites now festoon the walls, brightening up the blank white pockmarked cinder blocks behind the shelves containing the fleet of drones that replaced them.

The original copter cost about $5,000, though the truly autonomous models at that time cost more than $20,000. Learning to fly the drones was also easier. "It was much faster because they are pretty smart. They fly themselves," Dandois says, recalling that he started gathering consistent data for the first time within a month or two. "It was almost instant."

That said, "The first time we went out was a big crash, and I broke one."

The drones, after all, rely on radio waves to fly. Dandois flew the drone too close to an old watchtower, now covered in arrays to ensure continuous cell phone connections on similar radio frequencies. The drone passed into that active field of wireless communication and executed a perfect flip as it lost communication with its controller and fell from the sky like a stone. The copter crashed on the front lawn of a suburban home, the postmodern equivalent of hitting a ball into the neighbor's yard. Fortunately, no windows were broken, but the drone was dead. "That was a little traumatizing," says Dandois, who maintains a calm, even tone, punctuated with the occasional jolly chuckle, even when discussing the death of a $5,000 piece of equipment. "It was very, very dead."

Practice makes perfect, however, and over time Dandois and a team of undergrads worked out which cell towers to avoid, which drones worked best, which parts provided the most performance for the dollar. The German motors are more expensive but they also work better, though they come with restrictions on how the motor can be used due to German concerns about privacy. "They are built with an internal limiter, a kind of fence, so we can only tell it to fly within 250 meters from where it started," Dandois explains, or within sight of the person flying it. That limits the scale of the effort, though fortuitously, 250 meters by 250 meters is roughly the same scale as a megapixel in an image taken from space—the preferred cross-reference.

For the brain of the drone, Dandois, with the help of Gienow, settled on the Arduino microcontroller, the darling of the open hardware movement that allows the mechanically minded to freely share all the necessary knowledge for building usable electronics in the way that open software does for the wizards of computer code. This has enabled drone mini sailboats to ply the ocean, gathering info on marine life and cleaning up oil spills, and Japanese developers to program a personal Geiger counter, fashioning clip-on devices that turned an iPhone into a pocket radiation monitor after the meltdowns in Fukushima. As the T-shirt slogan goes: RESISTANCE IS FUTILE (IF LESS THAN ONE OHM).

The Arduino brain—and the open source software to direct it—replaced the built-in components, making the cobbled-together hybrid drone stronger, faster, and more robust. The entire working system—drone to nuts—can now be downloaded from the Ecosynth website and used by anyone anywhere in the world. There's still a lot of information to be mastered, Dandois admits. Then again, "Hobbyists do it."

Then there's the fact that thousands of pictures need to be knit together into one coherent image, and not just in the two dimensions of a flat photograph. The true purpose of Ecosynth is to create a picture of the Knoll that is rich in all three dimensions, as if the military had passed a laser over the forest and used its reflected light to build a complete holographic map. The challenge could quickly become what to do with all those pictures, all that information overload.

Drones like Dandois's are now everywhere, monitoring air pollution, delivering television imagery or packages, even killing people and stopping poaching. And those are the nonprofit drones. Entire businesses based on drones taking pictures for construction firms, mining companies, even farms (both solar and agricultural) have been launched, like

Skycatch. Farther up, beyond the sky, satellites have similarly proliferated. There are now thousands of satellites in orbit, ranging from school-bus-size spy satellites to tiny CubeSats that weigh 1.33 kilograms. Surveillance from space backs up surveillance from just above the treetops. The overview is now ubiquitous.

That has led to programs like Global Forest Watch, a spy system based on slightly out-of-date satellite pictures to show where tree cutting or burning is happening all over the world, whether deep in the Amazon rain forest, the clearing of the peat forests in Indonesia, or the logging of old woods in the Pacific Northwest. So far in the twenty-first century, the world has lost the equivalent of fifty soccer fields every minute of every day. That's a lot of trees and begins to give a scale of Earth's vastness and how human impacts have swelled to planetary scope. The Knoll is not much bigger than the soccer field the kids play on at the UMBC campus, and it is that stand's worth of trees, nearly 8,000 in all, that is cut down, burned, or lost to pine beetles and other pests every 1.2 seconds, the monitoring shows.

The satellites have gotten so good that every bit of the globe, all 510 million square kilometers of it, is parceled out into 30-meter-by-30-meter megapixels by, say, the Landsat series of satellites, now on iteration number eight. A similar system three hundred years ago would have watched the vast eastern forest of North America felled for farms and industry. Today's system watches that forest regrow, albeit in different form. The Knoll is a small part of that, part of a trend of restoration that has seen an area the size of Turkey covered in forests anew since the beginning of the twenty-first century.

There are 3 trillion trees in the world, according to the latest estimates. That's down 2 trillion or so, since farming appeared on the scene roughly 10,000 years ago. A combination of satellite images and the grueling work of folks like Boswell in more than 400,000 plots around the world enabled this more precise estimate, which is probably wrong but not by too many orders of magnitude. The number 3,000,000,000,000 looks like a lot, but it's not, given the clean air, clean water, and climate-change-combating ability forests provide. It's just 400 trees or so per person on the planet. At least there still are forests; the prairies and savannahs are mostly gone.

Satellite monitoring has already revealed that less famous forests are most at risk. Cambodia leads the world in deforestation, while the Congo—

the second-largest extant tropical rain forest in the world after the Amazon—dwindles, along with tropical rain forests in Sierra Leone, Guinea, and Liberia at an ever more rapid rate. Farm fields providing soy, beef, and even biofuels to consumers worldwide rapidly replace the drier, less famous forests of Paraguay and Uruguay. The movie to be made from satellite images over the last several decades would show the global forest shrinking from some edges, cut into in parallel paths like a fish's skeleton, carved into the woods from roads, or cleared out in proliferating big, square chunks.

It's not just forests. Satellites now reveal everything from shifting sea surface temperatures to the haziness—termed "optical depth" in the deathless jargon of scientists—of the atmosphere due to soot and smog. Satellites can even see the daily cirrus scars of airplane contrails and ship tracks back and forth across the Pacific and Atlantic.

Hic sunt dracones. Here be dragons, said the old maps. But the world is now mapped, and the remaining unseen places are few, mostly beneath the waves. No dragons have been found; the threat is clearly people. But that doesn't mean we understand what the complete map represents. The problem is sorting through all those views to find the ones that matter.

Like the paranoid, most people believe the world revolves around them. The idea of the Anthropocene could exacerbate that tendency. But just because people now play an outsize role on this planet does not mean we have grown up. The threat is us, the solution is in us. It takes a grown-up, rational species to recognize we are not alone and to provide for the protection of others, who have less control.

Global Forest Watch is a mash-up of environmental groups' concerns, the global governance of the United Nations Environment Programme and the U.S. Agency for International Development, the business reach of a Unilever, and the computing prowess of Google and its Earth Engine, which renders NASA Landsat satellite imagery into a zoomable globe (albeit one with some sensitive information blurred out). That same Google's computing power can turn Earth over time into a series of animated shorts of loss, such as the spreading sore of the tar sands strip mines in Canada and the attempts of human-controlled rivers to shrug off their bounds. It can also offer mini movies of lands conserved or restored.

People provide the ground truth that confirms or dispels what the satellites reveal. Ellis's team is not the only one using drones. Researchers at Wake Forest have hooked up an octocopter with sensors that capture infrared and visible light at the centimeter or even millimeter scale, watching a forest breathe up close and personal in a way (currently) impossible from space.

"Do you still fly the kites?" I ask Dandois.

He leans back in his office chair and strokes his beard. "No, we hang them up on the walls."

"There was no love for the kite once you left it behind?"

"Actually, it was pretty fun. It did kind of pass the time." He says it almost as if the impractical kites were better, more fun, more engaged than simply setting a drone with a robot mind loose on the forest. It's the difference between being a data processor overseeing systems that deliver good but cold and hard data, and autonomy.

When Erle Ellis went to China as an English teacher at Nanjing Agricultural University in the 1990s, he came back with a belief that the world was irretrievably changed. Just like the agricultural scientist F. H. King before him, he was impressed with the long track record of fertility maintenance in China, though not so much as to try a rewrite of King's 1911 classic *Farmers of 40 Centuries, or Permanent Agriculture in China, Korea and Japan*. King brought a new gospel of soil conservation back to the United States from Asia in the early twentieth century, trying to replace a system that relied on farming a given patch of land until exhausted and then moving farther west in search of fertile land. "That's Anthropocene thinking right there," Ellis suggests on his porch. "There's not an endless frontier anymore."

In terms of yields, Chinese peasants put American family farmers to shame at the dawn of the twentieth century, producing the highest yields in the world at that time, despite using some of the most intensively and anciently farmed lands. The Chinese "ran out of easy land in prehistory almost," Ellis notes. Yet China at that time was also the land of famine (proving the Nobel Prize–winning economist Amartya Sen's argument before he made it several decades later that famine was a result of distribution, not farming productivity).

Ellis's goal had been to learn and bring back some of that ancient mastery to reform Western-style industrial agriculture, since it is how people farm the necessary food, feed, and fiber in this current age that will determine what landscapes will be left. He ended up in the Mashan area of the small city of Wuxi in Jiangsu Province. Mashan was once an island but was connected to the mainland with the detritus from fish farms during the Cultural Revolution.

Working there, Ellis became convinced that the difference between a good farmer and a bad farmer in China was mostly down to historical accident, also known as luck. Yes, vicious hard work was required, and every last thing was recycled, but the farmers who did best were those in places that provided what might be called natural fertilizers, like farmers in the Yangtze River Delta who benefited from all that fertile muck washed off farmlands farther west. "They get free soil," Ellis explains. "It keeps coming down the river."

That made the differences between East and West seem smaller, a function of artificial fertilizers versus natural appropriation. "There's really no magic secret," Ellis says, noting the tremendous physical effort in shifting soil from plot to plot or dredging waterways by hand year in and year out. "They had the highest productive yields in the world at that time," he notes, "but you had to work like a slave."

Every part of the landscape is used. Fuel wood comes from the forests on the unfarmable hills and mountains. The ashes from their fires are saved and put on the fields, accumulating phosphorus in the soils. It's a farmer's ecology, the ecology of perhaps the most important anthrome and the most destructive one from the nonhuman perspective.

The problem for the Chinese farmers was nitrogen. The traditional solution had been so-called green manures, plants grown specifically to put nitrogen in the soil, such as legumes like peas. Or they fed the remnants of soy oil production to hogs and spread their brown manure on the fields. "They didn't know it was nitrogen, but they knew that that was the way to improve your yields." Ellis then cites an old Chinese adage that runs something like: "Raise pigs, lose money, get rich."

Chemistry is the enemy. The same nitrogen that crops require leaks into the air and flows away with the rains. As a result, the Chinese spent a lot of time adjudicating disputes over who had access to a given portion of canal or river sediment or set up little outhouses on the side of the road in the hopes that some poor traveler might use them and contribute some night soil to the farmers' fields.

That's what, in Ellis's view, brought on revolution after revolution in China: more labor required to extract the same amount of food from the same amount of land. The worry is that the world as a whole could be headed there. "You are working harder to get the same thing. There is nothing good about it," Ellis tells me. "It's horrible, actually. Imagine it: Just to keep your job, you would have to work more hours, every week, for the next twenty years."

I tell him this sounds a lot like the business of journalism in the twenty-first century.

"Exactly. You just work harder and get the same thing." Or, in the case of journalism, maybe something worse. "It is better than catastrophe, but it is still not good."

Catastrophes such as famine don't come about because the land has been pushed beyond its environmental limits. It's societies being trapped in the old ways of doing things. "People don't change their technologies until they have to," Ellis argues with a kind of bemused smile. "Things get worse and worse with the current system and, actually, we are forced to try something else and everyone has to do this new thing. It is not necessarily better in every way, but it allows you to move on." To survive, the price is lost landscapes and their inhabitants.

Chinese farming may have kicked off the Anthropocene with the advent of rice cultivation around eleven millennia ago. The creation of artificial wetlands to grow a water-loving grass with starch-rich seeds meant a lot more methane in the atmosphere, which has led some to suggest that a boost in methane is responsible for the summery climate of the last 10,000 years or so. The rice paddies of Chuo-dun-shan in the Yangtze River Delta from 6,000 years ago look pretty much the same as the rice paddies outside Shanghai today. Or perhaps it was the domestication of sheep, pigs, and especially cows, whose belching is responsible for nearly a third of the methane in the atmosphere as a result of human activities today. Certainly, the Chinese have successfully lived the longest in a human-dominated world. There are few parts of China that have not felt the impact of people. Every inch of land has its assigned purpose, fields alternating with brick kilns, and sheep wandering among orderly rows of trees. The Chinese countryside shows a vision of a future still to come in its uneven distribution around the world. "It is a great crucible for the Anthropocene," Ellis says.

China today is a very different place from even the country Ellis worked in a couple of decades ago. The current intensification includes fertilizer and soil, and the waste of sewage. Modern sewerage is a great blessing in terms of health, but it has disrupted the circular economy that brought fertility back to China's fields for centuries via the nitrogen and phosphorus in night soil. One of the best pivots humanity could make at this point is to turn away from wasting freshwater by mixing it with feces and urine and then funneling that mix into the oceans. Those vital elements need to find a way back to agriculture.

Ellis's thinking is a twist on a strand of thought that has often been called techno-optimism, the idea that more people means more ideas, means faster progress and thus the avoidance of catastrophes like famines or the end of oil. Instead it's a cycle of intensification to the point of diminishing returns that then force a shift to a whole new way of farming, for example.

And that's where we may be headed with synthetic nitrogen. In 1909, Fritz Haber showed how to transform the nitrogen that makes up more than 70 percent of the air we breathe into the kind of nitrogen plants can use as food. As a result, more than half the people alive today incorporate—have bodies that are literally made from—synthetic nitrogen. And farmers can grow more crops on less land.

Today even the lowliest Chinese peasant employs synthetic fertilizers and pesticides, a transformation of traditional farming techniques in recent decades of world historical proportions. Ellis recognized that the Chinese had imported Western-style agriculture based on manufactured nitrogen and made it over with Chinese characteristics. So he shifted his research focus to see where in the world the historical—densely populated villages doing subsistence farming—still existed, if not in China. "I set out to map it," he says, and from there he found the anthromes—twenty-one different ecosystems set up by people, for people, even if that includes simply leaving a place alone.

Studying humans and the ecosystems we make made him an outlier among ecologists, a group of people for whom people are largely anathema. So Ellis had little buttons made up with I ♥ HUMANS for the human ecology section he helped start in the Ecological Society of America.

People are the world's most (successful) invasive species, yet ecologists, in large part, do not study us, though there are now others, like Ellis as well as urban ecologists, who, by default, are dealing with people stuff. And people stuff now covers most of the arable land on the planet, even penetrating into the inhospitable climes of the tundra and desert. If a landscape hasn't been directly shaped by humans—whether a city, village, farm, or wildlife refuge—then it's been indirectly shaped, either by past use or the spillover from other ecosystem changes, whether acid rain or global warming.

What is wild, anyway? One thought might be Antarctica, which is still a place that will kill an unprepared person who steps outside—in other words, the deadly, or what's beyond our powers to dictate and must be respected. Or there's the poet Gary Snyder's evocation: " 'Wild' is a process, as it happens outside of human agency. As far as science can reach, it will never get to the bottom of it, because mind, imagination, digestion, breathing, dreaming, loving, and both birth and death are all part of the wild. There will never be an Anthropocene."

A simpler definition of "wild" might be "self-willed." And "nature" is defined by the *Oxford English Dictionary* as "the phenomena of the physical world collectively, . . . as opposed to humans and human creations." Ellis and others argue that opposition might be an illusion. In his view, humans are in a dance with our fellows in the natural world, influencing evolution in one place, sustaining it in another, not destroying but continuing a long-term relationship thousands of years old and changing every day. The idea of nature as something apart has deep roots in Western philosophy, stretching perhaps all the way back to the Garden of Eden as paradise, and real appeal for those troubled by aspects of the dominant civilization. But this view of nature also homogenizes humanity, obscuring the who in the we that is destroying or despoiling nature as well as the who that suffers from others' choices. It also fails to recognize that human nature is interdependent and the fate of that which is not human depends on how humans choose to lead their lives. The invention of the idea of wilderness is also the invention of its counterpoint: environmental degradation.

People's collective actions have changed the world. In China, elephants once roamed across much of the country but are now confined to wild corners of just one province in the far southwest. That's because people slowly but surely spread out to take all the land in China, leaving no room for other big animals. It's a story that's been repeated the world over.

In high school, Stephen Gienow was all about robots as part of Team REX 1727. Competitive robotics is not as aggressive as battle bots, where robots go wheel to tank tread in a chain-saw-versus-pneumatic-club battle, but the faux sports require real hardware skills. Think of it as football for 120-pound five-foot-tall robots, if the football players were controlled by three scrawny kids off to the side arguing furiously about what to do on each and every play.

Four years of this gave Gienow expertise in strategy (When playing soccer with robots, is it better to focus on scoring against the other team or guarding your own goal?), time management (Sport is announced and within six weeks high schoolers must have assembled their robot for shipping to the competition) and, obviously, problem solving. Plus, he had two years of actually operating the robot competitively.

Take the 2012 competition, featuring a game called Breakaway, kind of like soccer for robots. The field featured two half-a-meter-tall humps that the robots needed to be able to drive up and over without tipping over. The humps featured two little tunnels to allow minirobots, for teams that opted in that direction, to move through. At either end is a goal, and if and when the ball goes in the goal, it is carried up by a return track and dropped back in the middle of the field. The match lasts three minutes, and the team with the most goals wins. It would be hard to get further away from most ecologists' experience of nature. "That's where I got started tinkering with metal nuts and bolts, drilling, cutting things," Gienow recalls, in the clipped tone of an engineer. "That really got me inspired to do engineering."

But in college, Gienow could find only hobbyists interested in robotics, like the Inventors Club. "It's basically the build-cool-stuff club," he says. Through that club he found out that Ellis's group and, in particular, Dandois, were monkeying around with drones. "Flying things are like my favorite things, so of course I joined," he says. "It's like getting paid to do a hobby. It's really awesome."

That said, Gienow had never flown any kind of a remote-controlled helicopter prior to joining Dandois's effort, largely due to the expense. So he joined a program in search of a working robot that flies. Gienow's mission was to "try all these copter combinations and try to find one that works." To his credit—and with the help of a large supply of spare parts—he did just that, creating the hybrid copter with an Arduino brain the group (and Ecosynth devotees) now relies on exclusively. "Honestly, the whole time I've been at the program, my huge overarching goal was to make it so I'm not needed," he observes dispassionately.

That involves not only settling on the right hardware but also good programming, not typically a tool in the skill set of your average mechanical engineer in training. That starts with a "mission plan," basically the shape of the tree plot to be surveyed transected with flights to give full coverage. Next there's the "checklist" of necessary parts: extra batteries, rotors, wings. And finally there is the army of software developers who work with the Arduino, writing free code. With the latest iteration of the free code, Gienow thinks he can achieve the dream of ecologists who want to make drone surveys fail-safe: autonomous takeoff, flight, and landing.

"I would think that as somebody who loves to fly things, you wouldn't necessarily want to do that," I say as he prepares for another demonstration flight in the casual engineer's uniform of a red polo shirt tucked into conspicuously unwrinkled khaki shorts.

"It's not the physical flying around. Of course, that's fun," he admits, with no change in tone as his long fingers monkey with the drone and its controller. "I've been known to go up for a practice flight and hotdogging around one of the fields, saying, 'Oh yeah, the copter still works. Good.'"

But that's in an open field. For surveys, whether from within the Knoll or deep in the Panamanian rain forest, the drone has to navigate up through the canopy of trees. And that is no fun.

"I do not like launching through the canopy," Gienow says, a tinge of anger barely discernible in his otherwise monotonic speech. "It makes me very nervous because there's not much margin for error and there's a lot more pressure on you the pilot." Gienow should know. He once crashed the drone by flying it directly into a tree in the SERC forest, and maybe the Anthropocene will be distinguished by a thin layer of lost small technology—drones, cell phones, and the like. "My least favorite crashes are the ones that are completely due to human error."

Ellis's father, a D.C.-area journalist for *Time* and *Life* who covered the space program between the Mercury and Gemini missions, had a farm in western Maryland and raised Black Angus cattle back in the 1960s, when Erle was a kid. Young Ellis spent every weekend and all summer doing the backbreaking work required of a rancher—mucking the stalls, birthing calves, and other strenuous pursuits. "Nowadays, it's more or less a place to grow invasive species," Ellis says.

He's turned it into a rambunctious garden, and it was his interest in gardening that led him to plant physiology, his first scientific pursuit. It was also on the farm—this was the 1960s, after all—that young Ellis came into contact with the organic-farming practices, what Rodale called regenerative farming, that eventually led him to China.

Ellis's childhood dream, however, was to be an astronaut, entranced by the Apollo program to the moon, the idea of living in space, and the space helmets and other cool paraphernalia his father brought back from jaunts to Cape Canaveral to cover the space race. Erle thought he would be the agricultural specialist in a Mars colony. But by 1972, Apollo was gone, and by the 1980s, when Ellis was finally studying agriculture, so were dreams of a Mars colony.

"I had kind of an epiphany," Ellis explains to me, as usual on his porch. "This idea that if we can sustain ourselves without [this] planet, then we will be able to persist forever, or whatever. I don't know. I don't know what exactly. The ultimate sustainability idea, that we don't need nature at all." He pauses, mulling that moment when he came to embrace humanity's potential to not destroy everything on the home world. "We don't have to face that existential crisis. We can live here on Earth perfectly well. We don't need to have space colonies. Maybe we can have them, but that is not necessary."

This planet is the spaceship Ellis could work on, he realized. But when it comes to spaceship Earth, we have an observer problem. First, some members of the crew of the spaceship have particular priorities—think popular ecosystems like the coastal redwoods or any national park— that others don't share. The tall-grass prairies of the Midwest that have entirely disappeared, the original grassland gone, along with the way of life of the people who prospered in that landscape thanks to one piece of technology: the steel moldboard plow that cuts furrows and then turns the soil. Rain didn't follow the plow but corn and wheat surely did, farm fields replacing fields of grass swaying in the breeze.

Then there's the fact that ecologists have focused primarily on the most accessible ecosystems, not the most representative. This problem is not confined to ecology, naturally; it is true of many sciences, such as archaeology and paleontology. Our understanding of how a tropical forest works relies on only three tropical stations. In the world of ecology, Costa Rica and Panama stand in for tropical biomes and anthromes the world over. How representative is the isthmus of Panama for the rest of the world? Or even the rest of Central America? If it's not, then, as Ellis puts it, "you really have a lot of knowledge about one kind of condition, and very little knowledge about the world."

Take, for example, Barro Colorado, an island in the Panama Canal. The island is there because of a decision that people made. While digging the canal, a dam was built, which created a lake, which caused the water to rise and cover a lot of land except mountaintops. Thus, mountaintops became islands. Ecologists have been studying the effects of that decision on the flora and fauna ever since. One 50-hectare plot has been intensively studied since 1946, and biologists and ecologists have seen big predators disappear on the island fragment of what was once a montane forest while some smaller critters prospered. And this one mountaintop island in the Panama Canal has informed how ecologists think about the entire world as humans remake it. But is Barro Colorado really similar to anything else, other than other man-made islands of Panama?

The same could be said of Costa Rica, forest fragments in the Brazilian Amazon, or any other favorite ecology haunt. What we know of the Congo rain forest is often extrapolated from the Amazon, which is on another continent and dictated by different ecological and human circumstances. The Congo is not small—roughly 180 million hectares undergoing its own transformation, a kind of gaping hole or oversight in our knowledge of how the world works.

One way to begin to solve this problem is to get above to have a look, go into space. But we have another observation problem: The observations of the people who actually live in and around the Congo rain forest are rarely taken into account. Much cannot be seen from so far away in space, like the fine-grained functioning details of a forest. Much of that might be revealed by the people who have lived most intimately with that forest, even by the words of their language, a store of knowledge often ignored. In the interim, what's needed is a good sample from pretty much everywhere so we know what the world is like before we irrevocably change it again in the Anthropocene.

And that's where the drones might come in.

Ariane de Bremond, Ellis's wife, thinks her son Ryan is funny, especially when it comes to belaying. Ryan is learning how to climb on the walls of an indoor gym, in particular learning how to belay—securing a rope for another climber with your own strength to ensure if he or she falls, it's not far. "Are they strong enough?" I ask during my visit.

"Yes, they are," she says. "They are scared because they don't trust each other as much as they trust grown-ups."

And that's because the instructor has scared the kids with the dangers of learning to climb in a gym. The warnings include a broken leg, a broken back, paralysis for life. "The gym is a bad place to learn," cautions Ellis, and de Bremond agrees. "It is, because it feels so controlled, but it is not, really."

There is no gym to practice managing ecology in the modern era. Extinction is forever, as are range shifts, climate change, and other effects of civilization's "progress." And then there's the issue of trusting one another, which is crucial to any idea of a better Anthropocene. Yet the motto of the international community is summed up best as: "Distrust, and verify."

People are messing about in the real world, without any belay or safety net. On the one hand, that's good, because we should recognize that what we are doing is an entirely uncontrolled series of experiments. On the other hand, it's bad because there is no second chance to fix mistakes. When the Knoll is clear-cut or the mammoths are gone, the world is forever changed.

The problem might be the one that de Bremond has identified in her own work as a social scientist. Social scientists wait for that "output," that changed world, before offering any decisions or attempting to get people to change behavior, even using some of the tools or techniques for making social change. At present, it seems the only people with the money to even study the problem are the Europeans interested in global governance, a bogeyman to many Americans, even if many of these scientists do not accept that global government is the answer to global governance of global problems like climate change and mass extinction.

Ryan is an energetic kid who likes to tear things apart or blow them up in video games. The house is cluttered with the science books and posters of his parents, but also the spare bits of Ryan's drone fleet, some of the cast-offs of his father's work. "I have like fifteen things that connect to one thing, but nothing that connects to each other," Ryan complains. Plus the charger doesn't fit the battery.

This is the trouble with bespoke drone solutions. There are so many parts and pieces that after a while one can never be 100 percent sure which one goes with which set, a problem familiar to anyone who has ever built and rebuilt a Lego set. This is also the fundamental insight of some ecologists, perhaps best phrased by the legendary American ecologist Aldo Leopold: "To keep every cog and wheel is the first precaution of intelligent tinkering."

Yet when it comes to the cogs and wheels of an ecosystem, whether plant, animal, microbe, or nematode, we often don't even have a complete list of all the parts. "If the biota, in the course of aeons, has built something we like but do not understand, then who but a fool would discard seemingly useless parts?" Leopold asked. We may yet prove ourselves such fools.

"It isn't scientists' planet, you know?" Ellis tells me. "Maybe a lot of this stuff people want to do it, maybe a lot of people would give up wild forests." That's why Ellis has started teaching landscape architects at Harvard, as part of his cross-disciplinary outreach. From farming to ecology to landscape design, it's the art of the countryside. "Some of the most beautiful places on Earth were never planned," he says," yet they were completely created by people, or at least mostly created by people, like a Tuscan village."

Think of it as the accretion of human impact over time, shaping the landscape to suit our needs and, if we like, the plants and animals we deign to care about. For some, that means completely redesigning the landscape. For others, it means working with what's already there—say, the woodlot that became the Knoll—and building a landscape that is functional for humans, whether aesthetically or otherwise, around it. "The problem is, most of the bad stuff is hooked to good stuff," Ellis notes. "It's not a simple thing. You stop the bad stuff, you stop the good stuff too."

For Ellis, the Anthropocene isn't an epoch; it is a process, a way of reimagining, reenchanting the world—and even though he serves on Zalasiewicz's working group, he's not sure there is any need for an official geologic designation. For Ellis, the power is in the idea. Much as the Renaissance and the Enlightenment changed the way people thought, the Anthropocene could shape thinking going forward. We can rethink the way the world works, but only if we want to.

The new norm, even for ecologists, is aggressive intervention, whether in the form of rewilding or extermination, though the goals are conservative—to restore whatever was there before in some imagined, prelapsarian time. "Ecologists will bulldoze a place to get rid of an invasive [species]", Ellis notes.

More prosaically, tidal estuary marshes and islands got transformed slowly, over time, with accretion, into the capital city of the world, New York, in a way that renders the future unimaginable from the past. The ecological norm for some humans these days might be the air-conditioned office or home—a kind of hotel world laid over the top of other, older worlds, where people still scrape by in penury.

"We are not starting from scratch," Ellis observes. In other words, a prelapsarian time does not exist. Or if it did, it may be so far gone that it is unobtainable. The oldest Anthropocene may be an age of fire, the human mastery of burning to shape landscapes segueing after thousands of years into the comprehensive combustion of this planet's storehouse of fossil fuels. Perhaps a better name for the current epoch might be the Pyrocene.

Humans have always lived in environments shaped by humans, even when just a new species, small in number, freshly emerged out of Africa. "It is like beavers have never lived in environments without beaver dams," Ellis notes.

We live in a world of novel ecosystems, like the Knoll, where species are thrown together and learn to live and, sometimes, even thrive. In fact, these novel assemblages of plants and animals sometimes thrive so much that it may be impossible for them to tolerate a return to some original condition. And even if the original conditions could be restored, how could those conditions endure the shift in climate that is upon us? The Maryland of today is different from the Maryland of yesteryear or the Maryland of the next century. The process of becoming is eternal.

Take the Amazon rain forest, sometimes known as the lungs of the world (though it's really the plankton of the sea that produce the bulk of the world's oxygen). Will the Amazon persist, or will it be cleared to make way for farms and ranches? Or will it be cleared only to grow back, as with the eastern forest of the United States, or as may have happened before in this rain-forest basin when the Europeans arrived and killed off its human inhabitants through disease. The lost cities of the Amazon suggest the forest has not been left to its own will for millennia. "It's legitimate to say that the entire Amazon is not going to remain pristine, that's just not a legitimate plan," Ellis opines. "But the idea that we wouldn't consider keeping a lot of it wild? That would also be a foolish thing to do, not to think about that."

The danger of Ellis's way of thinking is complacency, that what we have is the best we can do and we can't really save anything, anyway. If there is no natural and there's no native anymore, then the world becomes whatever we can make happen and whatever we want. You end up in a world where anything goes. The rich among us are in the position once reserved for the Roman emperors, reshaping the landscape to suit their needs, just as Nero dammed the Aniene River to create three lakes under the patio of one of his villas in the mountains outside Rome. Or Seattleites leveled a hill to make new dry land out of 1,000 hectares of tidal marshes.

But nature is neither as fragile nor as robust as people think, which points up the flaws in human thinking. "To manage, you have to have some goals, so you have to have something you want to produce," Ellis says, whether that's to leave it all alone and let it prosper (the traditional ecology answer) or to intervene mightily. "A lot of ecologists say they have the answer. Just do nothing, leave it alone, and it will be fine is a common answer, but I don't believe in it myself." On the other hand, "It is quite possible that the best thing is to do nothing, just to let it unfold."

Fortunately for the drone, this day is not too windy. The smell of fresh-cut grass rises up from the ground while the cries of children playing almost drown out the closer whooshing whisper of the hexacopter as it powers up. And with hardly a signal, after an unseen command from Gienow, it rises into the air, making a stately beeline 120 meters straight up.

As *Crimson-I* buzzes up, the copter proves a bit wobbly but makes steady progress. Soon enough the drone is hard to track 250 meters up. Once at altitude, it moves in a more or less orderly fashion, dwindling in the near distance, toward the Knoll. Shading my eyes from the morning sunlight, I can see it fly a slow square through the sky while even higher up an airplane etches its contrail near the stratosphere. The cicadas twitter and fall silent, then twitter anew. The square flight pattern allows for the all-important grid of pictures, the camera snapping a new frame every two seconds or so during the twenty-minute flight. The pictures are to prove that the drone can help save the forests—and everything that relies on the trees therein.

The drone slowly wends its way back, sinking out of the sky as part of its auto-landing sequence. Its buzz slowly drowns out the cicadas. "When it gets to about ten meters, it slows down for landing," Gienow says. "When it hits the ground I'm going to—"

"Look at that," Dandois says excitedly.

"Awesome," Ellis says, clapping. "Good work."

Dandois stoops down and harvests the camera from its plastic casing. The flight-to-crash ratio continues to improve; an unexpected landing, sometimes catastrophic, now happens maybe every one in forty flights, maybe even one in fifty. That's not thanks only to the skill of pilots like Gienow, it's also the skill embedded in his flight programs.

Even the occasional crashes mean less now because the drones have become radically cheaper. From the beginning, with a first few bumpy flights, it became clear that the drone pictures knit together by software could deliver nearly the same (in some ways better) clarity as an overflight in an actual plane that uses lasers bouncing back off the canopy—a kind of laser-based radar known as lidar ("light" + "radar"). Now with hundreds of successful drone flights, the pictures give the heights of the individual trees along with size, which tells a scientist how much CO_2 is stored in all that woody, leafy mass, among other things. The pictures show the forest in full color, revealing subtle differences in shading and shade. And the ability to routinely fly over a given patch—whether in Barro Colorado or, maybe someday, the Congo rain forest—means scientists can watch the forest change over time, like the gentle planetary breathing of the seasons in a temperate forest like the Knoll. For example, Dandois now knows for sure that the tulip poplar leads the leafing charge in spring. There is no other cheap way to get that information.

Dandois and Boswell made a recent foray to Costa Rica, four days of up to eight drone flights a day measuring little 250-square-meter plots of restored forests to see how planting patterns—rows versus clusters—affect the regrowth, up all night charging lithium batteries and contracting a fine case of dengue that slows Dandois's chats with me. Dengue or not, the drones can go out every week, week after week, a kind of surveying impossible with the traditional methods. "We flew twenty-five flights, up to eight a day at one point, just flitting around the countryside," Dandois recalls. "It was very fun."

"Until you got dengue," I add.

"Until I got dengue, but that was after all the work was done I started coming down with symptoms."

"Perfect."

Autonomous flying eliminates the risk of dengue, as well as frees up ecologists and other interested parties to start gathering information on many more parts of the world. There's a whole new taxonomy of robots doing this surveying, from ever aloft solar-powered drones to gliders that plumb the depths of the ocean, surfacing to transmit data to satellites, like dolphins breaking the surface to breathe. As Ellis asks, "If it works on a cell phone, who can't do it?"

Google is already hard at work on Project Tango, an effort to transform cell phones into walking sensors for three-dimensional mapping, among other things, and old phones are affixed to trees to warn of illegal logging. Artificial intelligence running on Facebook's computers knit together new, more complete, more current maps of the world from satellite photos. But it's really about the people, going out and doing the work because they want to. The strategies are simple: management, technology and, *pace* drones, surveillance, like pigeons outfitted with sensors to detect air pollution. Drones can even be used to replant trees by dropping seeds.

There is a preliminary answer to what we supposedly want, enshrined in international forums and known as the Nagoya targets for conservation: 17 percent of the land and 10 percent of the coast and oceans protected by the end of this decade. A total of 25 million square kilometers of self-willed lands—an area roughly 1.5 times the size of Russia—and 36 million square kilometers (that's 10 percent) of the ocean, an area nearly as large as the entire surface of the moon. That may not be enough; some conservationists, like E. O. Wilson, are calling on us to set aside half the planet for plants, animals (besides us), insects, fungi, and microbes. Meanwhile, trillions of dollars are lost as bad farming practices clear forests, lose irreplaceable soil, and speed the spread of deserts.

"Those without the resources will suffer, and there is nothing new about this," Ellis says. "It's as old as the pharaohs." Maybe even older. Mother Nature is the font of all resources, yet she has no resources of her own. She has to count on us. Which is why efforts to hack the forest—whether the jungles of Peru or Ellis's Knoll—proliferate in this new epoch.

"Yes, the Anthropocene is going to be full of little tricks and turns," Ellis admits. "When your powers become greater, you have to deal with them whether you like them or not." Like the power of life itself.

Chapter 4

Big Death

We have created a Star Wars civilization,
with Stone Age emotions, medieval institutions, and god-like technology.
—E. O. WILSON, *THE SOCIAL CONQUEST OF EARTH*

North Dakota is not known for its pigeons. Or forests, for that matter. The state bird is the western meadowlark, a mellifluous blackbird (though it looks more yellow) often seen singing on fence posts. Such posts substitute for trees in much of North Dakota. The state is primarily covered in what was once short grass prairie but is now mostly farms embedded in a human-made grassland, exceptions being the Badlands and a swath of tundra forest in the far north near Canada.

Yet it was near Williston, the heart of western North Dakota's new boom-and-bust oil patch, in a log house on a bluff overlooking the Badlands and the Missouri River, that Ben Novak first fell in love with *Ectopistes migratorius*. For those unfamiliar with the Linnaean scheme for naming species in Latin—and the naming of things is a serious business—that means the passenger pigeon, a bird that rarely graced this region, if ever. In this latter-day patch of the Wild West, complete with (black) gold rush, the fifth-generation North Dakotan spent a self-described lonely and frustrated childhood and adolescence attending classes in a one-hundred-student schoolhouse in the tiny town of Alexander.

Novak comes from an animal family. He was not just surrounded by cats, horses, and chickens growing up but also taken birding by his farmer and auto mechanic grandfather most weekends. This was a man who "loved birds intensely," as Novak recalls, including the 250 or so canaries he kept in his house. In fact, Grandpa Novak set up a bird feeder in the line of sight of a telescope so he could concentrate on auricular feathers, those that cover a bird's ears. "You could never go wrong buying him anything bird-related," Novak says. The same might be said of his grandson, who raised doves, finches, and even pigeons as a child, including his first pigeon, whom he named Baldur, after the most beautiful and pale Norse god. "I name them only when they exhibit a personality," he says, though what bird personality matches up with the Viking deity of light, killed by a mistletoe projectile to herald the end of the world, is harder to explain.

In those days, Williston may have lacked excitement, but that is no longer the case. It has been a classic western resource rush, only in the Internet era and led by roughnecks with cell phones. Rents have been as high as in Manhattan or San Francisco, and Williston is the kind of town so booming that its Walmart could not keep the store shelves stocked, necessitating several hours of driving to the next nearest one in Minot or Dickinson, to the east and south. The town lives and dies by oil prices since the small city serves as the unofficial capital of the Bakken bonanza, a fresh flood of oil released by fracking that has reshaped global geopolitics. Locally, that has meant mushrooming trailer parks (known as "man camps"), millions of dollars' worth of planned construction in a countryside now dotted with drilling rigs, and gas flares that light the night, so many, in fact, that the state is visible from space during the dark hours festooned with blazing fires. The Walmart hires folks at more than $17 per hour, nearly double the national average, just to compete with the lure of oil field incomes.

This boom and bust has wreaked havoc on what was a small-town community, centered around the school that Novak attended, the local hospital, and other social services. All of them were overstrained, including the local government, which both tried to corral the boom and prepare for when the current bust inevitably arrived. Emergency services nearly broke under the strain. Novak's mother, an obstetrics nurse, witnessed the ER at her hospital, with only six beds, treat fifteen hundred patients a month. That included everything from oil field accidents and burns to knife fights in bars, blood spilling everywhere as if pigs had just been slaughtered in the medical facility.

Once, just a mile from Novak's childhood home—hand-built of logs by his father—a new well blew out and then blew up, becoming a raging inferno. The metal of the towering oil derrick turned red with heat, then crumpled and fell amid the jet of flame and inky smoke. No part of one man working that ill-fated rig was ever found, but another, who made it to the emergency room with Novak's father's help, had cooked fingers, swelled up like some horrible hot dogs, which then crumbled off the bone. "That's what comes with filling up your car," Novak observes of life in the current phase of the Anthropocene.

It was different for Novak growing up, just before the boom. He could indulge his passion for animals and science in the uncrowded school. Truth be told, the passenger pigeon was not his first extinct bird love. That honor falls to the dodo, a waddling ground dweller eaten to death by the Dutch and newly arrived rats in a span of seventy or so years in the 1600s. It is probably the most famous extinction at the hands—really the mouths—of humanity, because it was the first one Europeans noticed.

But the dodo hid a secret. It was in fact an overstuffed, giant-size, flightless member of the pigeon family, as a judge from the North Dakota Pigeon Association told young Ben as he stood in front of his posters suggesting how the dodo might one day be cloned back to life. A distant ancestor, like the now rare pink pigeon, flew to the island of Mauritius, found the place congenial, and slowly evolved over eons to lose the ability to fly and to bulk up in size. What made life perfect for these obese pigeons in the absence of predators made them perfect prey for sailors looking for grub in the seventeenth century. Ben dreamed of bringing them back and put together a science fair project (division winner) about how to resurrect the species.

The dodo is too far gone, most likely. So one day Novak, a precocious but solemn thirteen, found himself in the back of a Walden's Bookstore at the mall. There lay the science book ghetto, and it was there that he found the National Audubon Society's *Speaking for Nature: A Century of Conservation*. Just forty or so pages into that book is a picture of a museum display of stuffed birds, the male resplendent with a burnt umber chest and bluish-gray feathers on his head and back next to a more demure mottled brown-and-gray female. The text notes that "at the beginning of the 19th century, there were perhaps 3 billion pigeons migrating north to nesting grounds in New England and the Great Lake States. Early settlers commonly described flocks so immense that they blotted out the sun." Some settlers saw the birds as ill-starred omens, and they were—for themselves. Because most settlers saw them as mobile and abundant food.

By the end of the nineteenth century, this bird, once perhaps the most abundant bird in the world with a population roughly the size of all populations of all kinds of birds in the United States today, was gone. Hunters enabled by the twin technologies of the telegraph and the train wiped out the passenger pigeon by traveling from site to site to supply markets hungry for meat in the burgeoning cities of eastern North America. On September 1, 1914, the last passenger pigeon—named Martha and living in the Cincinnati Zoo—was found dead on the floor of her cage. The species was gone.

"I could not imagine billions of birds," Novak recalls. "It captivated me, I was rocked by it, and I was so sad." In the same way that young Ben mourned the end of the dinosaurs, he mourned the end of the passenger pigeon. But it was a little bit different. After all, an asteroid wrought the dinosaurs' doom, while the dodo, the Tasmanian tiger, Steller's sea cow, and more in a list that so far tallies more than three hundred known species had all been helped along to extinction by something less extraterrestrial—*Homo sapiens.* "There were generations of human beings that had these animals, and they cheated and robbed me and the rest of my generation from ever having them, in such an irresponsible way," Novak says, real anger coloring his typically soft voice. "There are extinct animals that we know nothing about other than that they tasted good, because no one bothered to write anything down." Or even draw a picture.

That's not true in the case of the passenger pigeon, because people stuffed them and then stuffed them into museum drawers. And it is from those museum drawers that Novak now pulls them, samples the DNA from a toe or toe pad, and hopes to bring a kind of passenger pigeon back to life, a living, breathing animal that a solemn kid like himself, an animal person, can raise and love. If he—or anyone else—can do that, well "then some twelve-year-old, thirteen-year-old boy or girl out there never has to experience the devastation I did when I opened up that book. Instead, they get to read about the horrible extinction story and then they get to read about mankind's amazing triumph in correcting it." Only the travails of ancient DNA, genomics, and synthetic biology stand athwart that obsession.

*   *   *

Our world provides the setting of a great murder mystery: Where did all the big animals go? Although Earth is the scene of the crime for many deaths—it is estimated that 99 percent of organisms that have ever lived are now extinct—roughly 35,000 years ago something unique happened: Most of the big animals disappeared from almost every continent. Australia's giant wombats and kangaroos as well as the marsupial wolf died roughly 50,000 years ago, the first to fall. Over the next 10,000 years, most of the big mammals of Eurasia, including the vegetarian cave bear and the woolly mammoth, died out. By 35,000 years ago, the murderer, whoever or whatever it was, struck North America, which saw the last of its mastodons, saber-toothed tigers, and horses dwindle and disappear. Even the bison almost didn't make it, reduced to a remnant population that then burgeoned into the vast herds of historical record according to the story written in the species' genes. Those big animals that did survive lived in remote, inaccessible places like the blustery, barren plains of the Arctic favored by caribou and musk oxen.

The animal lineages of North America have even been divided into the time before humans and the time after: Circa 14,000 years ago, the type and number of land mammals changed so radically that it's similar to the divides between geologic epochs. Dubbed the Santarosean and Saintaugustinean for the geographic locales that show a particularly good example of the split, they divide neatly around 400 years ago, when domesticated cattle, pigs, sheep, and the like entered North America in large numbers. Santarosean derives from the Santa Rosa Islands off the coast of California. Humans arrived here when these islets were still attached to the mainland back in the late Pleistocene. The islands contain fossils of both land and marine mammals. But it is the first appearance of skeletons of *Homo sapiens* in North America that completes the designation. The Saintaugustinean is named for the oldest continuously settled city in North America—Saint Augustine, Florida—where Spanish settlers first brought domesticated horses back to the continent of their distant ancestors' origin. Those equine ancestors and their descendants were extinct, perhaps thanks to the humans of the Santarosean.

Big is bad for animals during extinction times. Big-bodied dinosaurs died out while theropods that had progressively shrunk over 50 million years survived as birds. Insects have proved the ultimate survivors, and people have tiny mouselike mammals to thank for our present position atop the food chain. Mobility helps too, the ability to leave an area that is no longer fit for life. But what if there is nowhere to move to or something blocks your way?

Who done it? The climate did change around this time, but not that much, getting slightly warmer but not nearly as warm as periods that such species had survived in the past. Was it some kind of a hypervirus, a disease so deadly that it killed all the big animals the world over—yet somehow spared humans, walruses, and other, smaller critters? Or maybe some kind of big meteorite struck North America, killing off the big animals just as a similar rock from space wiped out the dinosaurs? The mystery remains unsolved.

There is a woman trying to solve this mystery: Beth Shapiro and her team of eager young postdocs and undergrads in T-shirts and jeans, setting up the first full-suite ancient DNA laboratory in the New World at the University of California–Santa Cruz. Nestled among the redwoods, there sits a building so new when I visited that it didn't even have a name yet, where Shapiro's husband Richard "Ed" Green and his team puzzle out ancient DNA by amplifying it over and over again until they can begin to see the patterns within the As, the Cs, the Ts, and the Gs, the molecules adenine, cytosine, thymine, and guanine that constitute the genetic instruction set encoded in DNA—plus whatever has been introduced by contamination. The adjacent computer lab has a futon for naps, since the programmers in the "shark tank" have to puzzle over code for hours or even days at a time. Green is the one who devised the algorithm to map the Neanderthal genome. There's also a yellow BISON CROSSING warning sign, perhaps a signpost for the future.

Like archaeology and paleontology, this new field of paleogenomics is severely sample limited: She who has the bones, has the ancient DNA. Add to that the technology limitations—the machine will reel off DNA held in place by the primers, but that DNA may not be ancient DNA—and the pressure on any given human to become an asshole becomes ever more acute. Simply put, there are more scientists in the field than samples to work with and make careers with, hence scientific papers with ten or twenty or even sixty names on them.

Some part of the impish and combative former child journalist Shapiro is a paleogenomicist for the shits and giggles, I suspect. I am the beneficiary of a master class from her on the fauna of the Ice Age that preceded the long summer in which humans spread across the world. "During the interglacial periods you had more warm-adapted things like giant sloth, and camels, and the giant beaver, and things like that . . ." she explains.

"Giant beaver?" I ask.

"It is the funniest extinct animal," she admits, her trademark dazzling grin breaking out.

"I didn't realize there was a giant beaver."

She points to a whiteboard on the wall that holds a grid of Pleistocene big animals her team is working on and lists progress toward understanding each animal's genetics. Is a sample extracted from dug-up bone or carcass? Is the DNA preserved? Can it be sequenced? If the answer to all those questions is yes, then it goes into the ancient DNA genomics library, a giant book of extinct life.

"Not just a giant beaver, Rob's giant beaver."

"Who's Rob? How does he come to own the giant beaver?" I ask, though I never get a straight answer about this mysterious postdoc or researcher.

"We continue to work on the giant beaver, not because we have been able to get any DNA from any giant beavers but because it's hilarious and we love to be able to say, 'We are working on the giant beaver,'" she says with a straight face now, just the hint of a laugh in her voice.

"Yes, that makes sense," I admit.

"It was five feet tall," she adds, or roughly her height.

"Five feet tall. Wow, I wouldn't want to run into that giant beaver . . ."

"See? It's funny, isn't it?" And she smiles again.

What isn't funny is what happened to the giant beaver, and all the other big animals no longer roaming Earth. They no longer exist. And there is one suspect who shows up on the scene time and time again just as the big animals start their march toward extinction—us. *Homo sapiens* arrives in Australia by, at the latest, 40,000 years ago. It takes a bit longer for us to spread out across the world's biggest continent, Eurasia, but once we do, big animals soon disappear. There isn't much sign of us in North America 35,000 years ago when big animals started to decline, but archaeological sites keep being found that extend our arrival further and further back in time. So while correlation is not causation, the story happens again and again and again. People come, big animals go. We are the charismatic megafauna—and, really, what animal can be as charming and cute as a human child?—that kills other charismatic megafauna: mammoth, mastodon, saber-toothed tiger, giant beaver, and giant kangaroo. We're often not great for uncharismatic microfauna either, but we tend to notice the uncute less (so typical of the charismatic types).

We may not have done it directly, just an accomplice to some other cause, though we may have too, hunting big animals to death. In Australia it may have been the human penchant for setting fires to clear a landscape and flush game that changed habitats and reduced numbers. "We are just an animal," as Novak says, echoing Erle Ellis, and "every animal manipulates its habitat." And our manipulations have lately, in a geologic sense, gotten global in scope. Or maybe the tag team of mild climate change along with human pressures proved too much for the biggest animals like the giant beaver. If so, that's no laughing matter for this new epoch.

The sixth mass extinction in Earth's history is gathering pace, the climate is changing, and our species remains the prime suspect. The hunter is now the hunted.

I came to Santa Cruz for one purpose, and it wasn't the redwood groves, beaches, or pot. Up the hill from a world surfing reserve is Shapiro's lab, where they're trying to understand how to pull off this new trick called de-extinction. I came to find out why a young man from North Dakota had become so obsessed with the passenger pigeon that he has devoted his adult life so far to the perhaps quixotic pursuit of resurrecting it.

I first saw Novak on a stage in Washington, D.C. It was March 15—the ides of March and a "weird day" in the words of convener Stewart Brand, an icon of the counterculture of yesteryear and the cyberculture of the Bay Area today. We were gathered to hear about a premonition of news regarding the possibility of bringing back the dead. Novak wore an ensemble in dark brown, gray, tan, and silver that made him resemble a gunslinger, minus the guns and boots, and with a thin hipster beard, too well groomed for a real cowboy. His long dark hair flopped over his eyes despite the sides and back being shaved, a look popular among 1980s punks and Weimar Germans. He came to the offices of the National Geographic Society to tell a fifteen-minute tale of resurrection, a would-be Lazarus species. Before he or anyone else started talking, a big picture of a woolly mammoth flashed on the screen with a warning writ large: DON'T MAKE ME COME GET YOU. TURN OFF YOUR CELL PHONE.

That is the organizers' fervent wish—a world where the woolly mammoth could come get you and put you in your place, thanks to the proboscidean species resurrection at the hand of scientists and rewilding via conservationists. The mammoth is certainly the poster species for this effort, dubbed de-extinction, sold as a plush stuffed toy or used to advertise the Ringling Brothers circus when it comes to town full of fantasy.

The event was cohosted by TED, Technology, Entertainment and Design—a nonprofit "devoted to spreading ideas," according to its own PR. This one had an x, which stands for an independently organized series of talks, the indie rock of what is becoming an advertising industry for pop culture distillations of big ideas in business or science. "The x used to be self-organized, but it really now stands for TED multiplied," explained TED's founder and chief proselytizer Chris Anderson. The talks—or "Ideas Worth Spreading," in the lingo—have a long afterlife on the Internet (you can find them with a quick Google search).

There were haunting photographs of extinct species by National Geographic's stable of professional photographers and art including Isabella Kirkland's still lifes, teeming with evidence of species that have gone extinct at human hands. Her epic paintings capture at life size and full color the many species of plants and animals we have threatened or already made extinct. Turns out the eggs of the (extinct) great auk—a kind of North Atlantic penguin, if you will, also eaten to death by hungry sailors—were, each one, as unique as a snowflake. If butterflies are more your thing, there's the Xerces blue—iridescent with the color of its name as it fluttered from flower to flower—last seen in the 1940s, wiped out by the growth of the beautiful city by the bay known as San Francisco. Which begs the question: Which is more beautiful to human eyes, a butterfly or a city? As the eminent biologist David Ehrenfeld of Rutgers University—among the first to cry out for the preservation of the breadth and diversity of life on Earth and to be skeptical of human faith in our own technology—noted, "At this very moment, brave conservationists are risking their lives to protect dwindling groups of existing African forest elephants from heavily armed poachers, and here we're talking in this safe auditorium about bringing back the woolly mammoth?"

Well, yes. "Environmental destruction, loss of habitat, species relentlessly clicking off the map," Anderson said. He was following a slew of pictures of such soon to be clicked species that had flashed by on the big screen while the audience waited expectantly for the event to begin: the woodland caribou, Palos Verde blue butterfly, and Higgins' Eye pearly mussel, which seemed to make eye contact with the audience. "I'm sick of it. I'm sick of it!"

And that's what de-extinction is meant to counteract, this sense of impending doom, the idea that conservation has been a heroic effort to win a few battles while losing the war.

Novak had taken to the stage to talk about not bringing back the passenger pigeon. "No book and no museum collection can ever give you the majesty of what this animal once was," he told the audience, trying to impart some small part of his own obsession. This was no bumpkin from the sticks but a salesman whose polish came from not being polished and pure passion. "I want to bring them out of our imagination," he said, noting that it was his fervor that got him the job of passenger pigeon resurrectionist in chief, "and get them back into the wild."

This is not as ridiculous as it sounds in the cold light of YouTube viewing.

But why bring the passenger pigeon back? For Novak it appears to be the particulars of that bird: its color, size, shape, and even its odd behavior of dense flocks that made it so amenable to hunting and like a shitstorm as it passed overhead or roosted. For him, the passenger pigeon was the only fitting dance partner for the forest, capable of flitting en masse from outcrop of acorn to outcrop of chestnut.

So I arranged to meet Novak in a conference room on the floor above the main event. He slouched in his chair and did not make much eye contact, becoming animated only when I pressed him on the likelihood of any of this passenger pigeon resurrection actually happening. He had left a gig at the forensic genomicist Hendrik Poinar's lab in Ontario, where they are working on the mammoth's DNA, among other things, strictly to pursue the resurrection of the passenger pigeon full-time. "I've been given my dream job without having those three letters," P, h, and D, Novak notes. "There is no paperwork or tests, and I get paid more than a graduate student."

I said that after him I would be chatting with another eminent geneticist, George Church, who stands tall at nearly two meters in height, burly at roughly 110 kilograms, and sports a longish white beard. He looks like Leonardo da Vinci's picture of God and is perhaps the most prominent proponent of synthetic biology—the human manipulation of the very stuff of life, the genetic code—with the white robes replaced by the sober but rumpled suit of a seeming Boston Brahmin who is actually originally from Tampa Bay. Inspired by the 1964 World's Fair in Queens, famed for its Futurama depiction of the near future, he took up biology rather than redesigning cars or building robots. His fundamental idea was simple: Bring the tools and techniques of engineering to biology and unleash a world where cells could be remade to do people's bidding, whether for curing disease or churning out fuels to replace gasoline. In the lab where he oversees ninety students of varying degrees at Harvard Medical School, one experiment stands out: Asian elephant cells reconfigured to come closer to those of a woolly mammoth. As the longtime innovator says, like so many of his fellow entrepreneurs, "The best way to predict the future is to change it."

I explained to Novak that I found it hard to parse what specifically Church would do to bring back the mammoth. He commiserated. A few summer months spent in Church's lab among the revivalists there had been similarly unenlightening for him. Even the leading proponent of passenger pigeon de-extinction did not pretend to fully understand the manipulation of biology required to make it possible.

That's why Novak had headed out to Santa Cruz and Shapiro's lab, in search of more hospitable climes for his de-extinction science. After all, before any genetic code could be manipulated to make a passenger pigeon, the genetic code of the original passenger pigeons would have to be known. And what better place to do that than a newly built lab at the University of California, nestled among the redwoods? That's where one finds not just the scientific resources but also the inherent optimistic boosterism endemic to the end of the Western world. Then there were Shapiro and Green, the renowned paleogenomicists, prospectors for genetic code frozen in the carcasses of Pleistocene megafauna, such as horses and mastodons, who had already puzzled out the oldest confirmed DNA on record—ancient horse DNA 700,000 years old.

The only downside was that Shapiro did not think Novak had a real chance of bringing the passenger pigeon back. "Bringing back extinct species is probably not my end goal," Shapiro told me, on the sidelines of that same TEDx meeting as I set up a time to visit her in Santa Cruz. "If this is what it takes to inspire people, then more power to it."

So it was a marriage of convenience: funding and grunt labor to get the passenger pigeon genome sequenced for Shapiro and the chance to pursue the resurrection of the passenger pigeon for Novak and his funders, the Long Now Foundation. Long Now is one of the fonts of techno-optimism in California—exemplified by the canny old ecologist Stewart Brand and his wife, Ryan Phelan, who made a fortune in genetics as applied to medicine. California is where techno-optimism remains most virulent, though it is now endemic to the entire United States and perhaps the entire world of affluence, wherever those lucky humans happen to roost.

The moneyed folk of the Long Now Foundation, like Brian Eno, who gave the group its name, believe that modern humans are too short-sighted and so have focused their considerable fortunes and talents on building a clock that will tick for 10,000 years and an archive of the diversity of extant human languages, including computer ones, among other projects to foster long-term thinking. They also host long bets, or "enjoyable competitive predictions of interest to society," as the foundation's website has it. The group has a fetish for adding a 0 in front of the date, such as its founding in 01996, to avoid a computer code bug that will come into effect in 8,000 years or so, a kind of long-range Y10K problem. The long now encompasses the entirety of human civilization, stretching back to when the Pleistocene ice retreated and agriculture was invented, and stretches forward to what histories may be encompassed in the year 012000. Let's hope that will not be the entire duration of the Anthropocene.

Long Now hosts seminars at its own private salon/saloon, the Interval, where one can keep a bottle of specially distilled gin or whiskey, presumably for decades, waiting amid the elegant décor and a chalkboard-drawing robot for taking down long bets as well as another robot to pour bespoke gins of 87 billion variations all at the touch of a cell phone screen. Brand, Phelan, and their Long Now cohort are also funding the resurrection of the passenger pigeon—or at least an exploration of whether such a resurrection is possible. Phelan's fortune from DNA was partially invested in the initial meeting back in February 02012 to determine whether it was even feasible. Church was there, arguing for it, along with Shapiro and the man perhaps as passionate about the bird as

Novak—Joel Greenberg, one of the founders of Project Passenger Pigeon, who now denounces the effort.

There was enough interest, including donations from private individuals, $10,000 here, $100,000 there, that a full-scale Revive & Restore effort was launched—and the TEDx event was planned. The money was enough to get Novak to Santa Cruz, though Brand maintains that "our shallow pockets are now dry."

When I arrived in Santa Cruz, fresh from the plane ride to San Francisco and a drive through Silicon Valley, past the Apple campus in Cupertino, up and over the old logging trail, I could barely see the beach from my hotel perched high on the mountainside. A drive into town brought me right down to the coast and to the amusement park. I then got a bit tangled in the streets of downtown searching for the campus. But after driving through ritzier and ritzier neighborhoods the closer I got to my goal, I finally reached the former farm and Wild West trading post that marks one entrance to the campus. Here the administration buildings have the look of repurposed red barns and sit at the foot of golden rolling fields known as the Three Meadows. The dorms and other buildings are hidden up the hill, nestled alpine style among the redwoods. The roads up offer vistas of town, sea, and mountains.

I proceeded up the hill, watching the black and brown cattle graze the golden hillside while sailboats tacked across the blue bay waters, before disappearing into the redwoods and turning into a hulking cement parking garage. A quick stroll through the campus—including a detour into the foot-padded trails among the redwoods—led me to the postmodernist lab atop Science Hill. I called Novak to let him know I had arrived.

We met up just outside the new facility and headed a bit down Science Hill to the environs of old Thimann labs, redolent with the reek of decades of chemistry and unknown, ongoing hazardous experiments. But no DNA had ever been handled here, which made it perfect for the actual ancient DNA lab—no contamination except what the people brought in with them. Such contamination is all too common, and when Novak calls up Shapiro—"Hi, Beth, this is Ben. Novak. Pigeon guy."—to get permission to bring me in, it is not forthcoming. Who knows what contamination I could bring or where I have been? The field of ancient DNA is rife with spurious results, including DNA reputed to be millions of years old trapped in amber, thanks to similar contamination. In fact, a claim of the discovery of dinosaur DNA was disproved by showing that it was oddly similar to the random mutations contained in the genetic code assembled from some scientists' sperm.

So Novak suits up in plastic from head to foot in the vacuum ante-chamber. Before putting the gloves on top of gloves, he finishes a rapid-fire burst of texting. Then he slips on a pair of Crocs and a plastic cap and disappears into the chamber of ancient DNA secrets.

My window into this world is a small one, next to a rusty carousel. He graciously shows me his workspace and specimens through that one window onto the bone-grinding room, which must be kept sealed to prevent DNA wafting in on the breeze. This is a step in from the changing room but a step outside the room where the DNA is actually extracted and the protocols are most sacrosanct. He holds up tiny foot bones and feet while swathed in the protective gear of an Ebola worker. This get-up was to prevent his contaminating the bones rather than the other way around for health workers. Imagine making thousands of copies of good old ancient bison DNA, some of which gets onto your hands or hair (DNA is promis-cuous) and is then carried by you into the passenger pigeon lab to wreak havoc on scientific understanding. The worst part is that the scientists might not even know it is spurious DNA for months, years, or, in the worst case, maybe even never.

That's why they don't use the gleaming new labs for the real work. Geneticists, microbiologists, environmental toxicologists, developmen-tal biologists, biological physicists, cell biologists, molecular biologists, and even biomolecular engineers all share this space. "I'll check your pockets when you leave," one professor jokes to his graduate student, swiping him into the lab. UCSC may have missed out on the nuclear weapons science that brought fortune and glory to Berkeley and Caltech, but it does not plan to miss out on the biology boom. UCSC is gearing up to be one of the two big ancient DNA labs in the country, the other being at Harvard, though the evolutionary geneticist Svante Pääbo has built one in Leipzig that might one day churn out humanoids. But Santa Cruz offers much that the entirety of Germany cannot: the cool shade of the redwoods, the sweet smell of pine, and access to the California beach.

But what I really wanted to know was not why Novak spent time in the puke-green anteroom and its reverse vacuum putting on plastic from the lockers that line the wall, all so he could handle delicate old bones and try to prize the secrets of passenger pigeon life from them. I wanted to know why he was so obsessed with passenger pigeons that he made paintings of them and dreamed about them, and whether the story of the passenger pigeon might be the story of all life in the Anthropocene. "People like Ben, who are passionate, sometimes lose track of the details," says Matthias Stiller, a Shapiro postdoc who knows obsessive passion from time spent in the ancient human DNA lab of Pääbo, the mad Swede, where he wasn't allowed to touch the pipettes or even read their labels. "But without that passion it wouldn't happen."

In the fall of 1847, the residents of Hartford, Kentucky, held a pigeon massacre. The passenger pigeons had recently come to roost in the area and could be shot, poisoned, netted, or clubbed with ease. Already, the pigeon was becoming a rare sight on the East Coast and had retreated inland, though at some point in the early nineteenth century the passenger pigeon was probably the most abundant bird in the entire world. No wonder John James Audubon could recount in 1833 that he had seen a mile-wide flock of the fowl passing overhead for three straight days, blocking the sun.

This particular massacre and its attendant fricassee was witnessed by the French bon vivant Benedict Henry Révoil, who went on to publish a book on shooting and fishing in the rivers, prairies, and backwoods of North America. Whether it's for fur or feather or meat, our cruelty to our fellow animals knows few bounds. But even this epicurean hunter could see that the continent's most abundant bird would not long withstand this kind of massacre. "Everything leads to the belief that the pigeons, which cannot endure isolation and are forced to flee or to change their way of living according to the rate at which North America is populated by the European inflow, will simply end by disappearing from this continent, and, if the world does not end before a century, I will wager . . . that the amateur of ornithology will find no more wild pigeons, except those in Museums of Natural History," he wrote in a book that dealt with the pleasures to be found in the New World, not least gluttony.

Révoil won that wager. The drawers of natural history museums are exactly where Novak finds passenger pigeons today, 15,000 known specimens from mounts or fossils. There hasn't been one in the wild since the turn of the twentieth century, but pigeons from more than one hundred years ago roost in museum collections. Novak found his first in Chicago while on a mission to collect mastodon for ancient DNA specialist Poinar of McMaster University, a little flock of dead birds nestled in a drawer along with tags revealing the birds were shot in the midst of breeding season. But that was just the first of many: The largest cache he has found so far is seventeen dead birds from the same flock in a drawer in Denver.

The last living passenger pigeon was also the first passenger pigeon individual: Martha (named after Martha Washington). Her individuality spelled her doom, making it impossible for her to mate or even to properly construct a nest. She was the last, like Lonesome George, thought to be the last of the Pinta Island giant tortoises, now dead, or Benjamin, the last thylacine, better known as Tasmanian tigers, who lived long enough to be filmed pacing his cage in apparent distress. Benjamin's manner is all too familiar from the rare animals who pace the limits of confinement in zoos around the world today. The Anthropocene could yet end with all wildlife confined in a zoo.

Martha suffered that same fate. She lived all her life in captivity and ended her days in the care of zookeepers Salvator Stephans and his son Joseph. She had a mate, named (of course) George, for a time, but no offspring. Visitors described the birds calling to each other with the sound *see, see.* But George died on July 10, 1910, leaving Martha a widow. And so the last living representative of a highly social bird that found comfort in flocks blotting out the sun lived another four years in solitary confinement in a cage measuring just six by seven meters.

What is known with some precision is that the last passenger pigeon died some time on September 1, 1914, a hot and humid day in the upper Midwest, which is pretty precise for an extinction. No one knows when the last baiji, a Chinese river dolphin, died, and that's the most recent extinction on record. Depending on whose account you choose to believe, Martha could have died early in the afternoon or later that evening. And she could have been surrounded by mourning zookeepers or simply found dead on the floor of her cage. Regardless, when the sun rose on September 2, 1914, there were no more passenger pigeons in the whole wide world. A bird that has been described as a "living wind" flew no more.

Upon her death, her feathers and her corpse were shipped on ice to

Washington, D.C., to be stuffed and preserved. You can still see Martha today in the Smithsonian Institution, her plumage restored but not her vivacity.

The toe pads of her fellow museum specimens may help with that. Molecular ecologists have pieced together much of the passenger pigeon genome from such resources, mostly to fulfill some idle lunchtime curiosity about "how evil humans can be to wildlife," as recalled by lead researcher Hung Chih-Ming, now of Taiwan National University. Four male birds bagged in Indiana, Pennsylvania, and Minnesota and the genetic variation they carried suggested that passenger pigeons may have been what ecologists call an "outbreak species." That is, a species that is prone to huge explosions in population when the conditions are right, like locusts. The trouble for such species is that their numbers inevitably plummet with the attendant decline in available resources—for the passenger pigeon, forest mast of acorns, beechnuts and, back then, chestnuts—or an orgy of hunting from typical predators.

That in turn suggests that what Audubon and Révoil witnessed was such an outbreak, perhaps brought on by European elimination of the main competitor and predator of such passenger pigeons—the original Americans. But the same Europeans who spelled boom for the passenger pigeons also meant bust. Settlers cleared the pigeons' forest home, domestic pigs grew fat on the same mast, and hunters decimated the skittish bird, preventing it from roosting and mating. As Audubon noted, only the reduction of the Great Eastern Forest could fell the passenger pigeon, and that forest fell even faster than the great naturalist could have imagined.

Still, the genome suggests that passenger pigeons had survived massive reductions in numbers before, such as during the most recent Ice Age, and then recovered. De-extinction might then not need to create billions of the bird; hundreds or thousands could establish a viable wild population. Hung is skeptical, noting the considerable social issues—such as passenger pigeons destroying crops—that would attend any reintroduction. Just ask the ranchers who seethe at the reintroduction of the wolf out west, or the Englishmen who resist the return of the beaver for entirely irrational reasons, like the vegetarian mammals somehow depleting fish stocks. De-extinction "is interesting," Hung suggests. "But I personally do not think it is a worthy step to take for conservation."

Such skepticism does not deter young Novak. Pigeons are almost the perfect bird for resurrection, he notes, because pigeon fanciers have been breeding them into fantastic shapes, sizes, and colors for millennia. Even Charles Darwin saw the special merit of pigeons, breeding them to help gather evidence for his nascent theory of evolution and using them as a model organism to demonstrate facets of his new theory as laid out in *The Origin of Species.* In the very first chapter of that foundational text, he notes that breeders had created varieties, almost species, as different as tumblers, carriers, pouters, Jacobins, and fantails, but all descended from the common rock pigeon (*Columbia livia*). Nature might do the same through the long, careful work of natural selection, for which pigeons provide a handy example. Can a white pigeon long survive the predation of its hawk hunter in the wild? Yet Hung's research shows that the genetic secrets hidden in Martha's desiccated corpse may still bring some version of her back. And we can always paint the pigeons, like the fanciers do.

A variety of craft brews dug out of the backs of mini refrigerators—like ancient DNA preserved in permafrost—serve as lubrication for an impromptu seminar on the genetics of really old animals (and, to some extent, plants). The fridges also store, potentially, the genes of ancient horses or, one day perhaps, giant beavers.

The field of ancient DNA is, at its best, a form of time travel. Theoretically, one day paleogenomicists could actually watch genes at work over time and thereby learn to read the map of genetic change as genomes respond to external circumstances. Take the horse, domesticated for perhaps 4,000 years. How do the genes of a modern horse differ from those of the horse that Ed Green sequenced from 700,000 years ago? There are roughly 2.7 billion letters in the horse genome—which ones are important? One way to find out might be to put genes of an extinct animal into mice using all the novel techniques of genome editing and then see what happens. That would also open a way to figure out what errors there are in the sequencing and ultimate genome, then go back and correct them.

Then there is the idea of bringing this ancient life back: perhaps a dwarf mammoth to hold your newspaper or a small scimitar cat to open letters? All it would take is the genetic code, since code is destiny.

Assuming the code is in hand, there are multiple methods for potential de-extinction. First—and already successful, though all too briefly—there's cloning. You can either take a hollowed-out egg cell and dump the contents of a cell you want cloned within, as was used to resurrect the extinct species of Iberian mountain goat known as the bucardo, which lived for all of seven minutes. Or you can reprogram a skin cell with the proteins that dictate what DNA to copy or not, so-called transcription factors, to create a stem cell capable of becoming anything, including an entire organism. Korean scientists routinely use the technique to clone cows, dogs, and pigs, and hope to extend it to endangered animals like the Ethiopian wolf. But there are all kinds of struggles with these methods when it comes to de-extinction: ill-formed, enormous placentas, immune system failure—and all for reasons that remain mysterious.

And then there's CRISPR-Cas9, the technique overseen by George Church. The acronym stands for "clustered regularly interspaced short palindromic repeats." The repeats let the genetic machinery of a simple cell, say, *Escherichia coli*, the gut bug responsible for most cases of food poisoning, to recognize a viral intruder and dispatch an enzyme to kill it. Viruses are, after all, just bits of DNA or RNA code looking to hijack cellular machinery to make more of themselves. Instead, Cas9—for CRISPR-associated system number nine—attaches itself to specific stretches of viral DNA and cuts them, disabling the would-be pathogen.

Now geneticists can use this same rudimentary immune system to edit genomes, adding or subtracting genetic sequences at will. The technique can be used to tailor life itself, editing the parasite that causes malaria out of existence or inserting genetics for drought tolerance into key crops like wheat. Or it can be used to resurrect a pandemic virus, like the influenza that killed at least 20 million people in 1918, or perhaps to craft an extinct genome from a living one.

Already, Church and his clutch of revivalist students have tweaked genes from an Asian elephant to produce more hemoglobin, the protein in blood that carries oxygen. That's the first step on the long road to making a hairier elephant that can tolerate the cold, a living analog of the extinct woolly mammoth. "We can make a hybrid elephant with the best features of modern elephants and mammoths," Church says. Next up, organs.

But a whole animal is a different story, as the trial-and-error saga of cloning shows. "Development is like a fine watch. Changes have consequences," cloner in chief Robert Lanza of Advanced Cell Technology tells me. "You can't realistically change 10 percent or even 1 percent of a genome and have that go to term."

To get there, at least eight more, potentially Nobel Prize–worthy breakthroughs are necessary, and those are just the known hurdles. As it stands, there is no technology for editing an entire genome, only for individual sections, no synthesizing the genetics of complex organisms or wrapping them in the tightly packed DNA known as heterochromatin necessary to get them to function. Gene expression of ancient animals that no longer exist remains a mystery. And then there's the damage done to ancient DNA by the actual tools of genome reading, which can cause misreadings of code, especially at the ends of DNA fragments. And don't forget the problems of contamination. "It's a traveling-to-the-stars kind of thing," explains the Englishman Peter Heintzman, another young member of Shapiro's lab.

How many genomes have been entirely sequenced? None, not even that of modern *Homo sapiens*. Ancient DNA will likely never produce whole genomes, requiring multiple reads from multiple samples to even come close. The best may be the mammoth, whose genome, tirelessly puzzled out by laying seemingly overlapping sequences over one another until a three-quarters-complete genetic code is revealed. As Stewart Brand puts it, "We're not working with a full deck."

And even once that is done, the specific workings of each bit of, say, the 3.5 billion nucleotides in the human genome would need to be puzzled out. Get even one base pair wrong and the whole organism may be doomed. In fact, "natural" genomes may never be fully understood because they evolved, developing randomly. An annotated genome becomes the most fervently desired and sought for scientific outcome: a single genome, of any kind, really, in which every gene's purpose and function are known (or known to be irrelevant) and even the various things that happen to a genome—methylation, transcription factors, epigenetics, and more—are understood and predictable, more or less. That remains a long way off.

But the mitochondria are known, the powerhouse of every living cell with its manageable genome comprising just 16,500 nucleotides. Our seminar's conversation devolves into rhapsodizing about the benefits of various technologies in reading DNA, like labs on a chip, akin to more familiar conversations about iPhones and various apps. Born in East Germany, Matthias Stiller's family had to wait eighteen years just to acquire a car. Now he has access to the world's best sequencing machines. Thanks to them, the evolution of viruses can be watched, with National Institutes of Health funding and tractably small genetics (8,000 or so base pairs), for example. But is a virus even alive? Scientists still argue that one.

The field of ancient DNA is an incestuous one, being small and new. It is part of the radiation of a new scientific species, the ancient DNA specialist or paleogenomicist. At the same time, the paleogenomicist must fit into a landscape in which science is becoming more integrative, with cross-discipline collaboration becoming the norm as what were subspecialties collapse back into one.

The problem faced most squarely by those who dabble in ancient DNA is corruption—changes to the DNA not reflective of the genetic secrets of the animal or plant that bears them, but rather of the rigors of museum storage or permafrost freeze. To overcome that requires sheer numbers, as well as being able to distinguish among the random variation of individuals in a population. "Mammoth DNA is more like confetti that's been run over by a herd of mammoths in the rain," noted Shapiro at the de-extinction event in D.C., punctuating her metaphor by tossing confetti into the air onstage.

That means the challenge of ancient DNA is sample collection, like archaeology or paleontology, trudging up to the tundra to carve out frozen corpses from gold-mining sites or thaw zones. Santa Cruz could not be more different from the Arctic where Shapiro captures her "data." Then there are the amateur collectors, such as the Dutchman who trawls the North Sea to dig out fossils from drowned Doggerland—like Neanderthal skulls or hyena bones—and sells them on eBay. Forty percent of the land that was above water in the Pleistocene is now beneath the waves, hiding treasure troves of ancient bones and, perhaps, ancient DNA. For Novak, at least, it's slightly easier, being more a matter of rooting through museum collections, turning up mummified corpses of passenger pigeons left to desiccate in a drawer with even the contents of their stomachs intact, by a careless conservator.

As the nascent paleogenomicist Stiller puts it, "We are in the stamp collection business." Those stamps can include some pretty strange things, including Poinar extracting poo from sperm whales to sample the rich panoply of DNA in feces. As we saw in the first chapter, whales are biological pumps, recycling iron from the deep sea back to the surface. How long has this dance been going on, and to what effect? That is the next step in the field, to move away from simple stamp collection and toward watching evolution through time. That is, to look at the climate changes of the shift into and out of an Ice Age, starting with the Pleistocene. Did each species respond differently, or were there general trends? Did humans kill off everything they could reach to eat, or did climate, as it usually did in the more distant past, do the job? Most likely, it was a combination, the same combination people are reengineering right now—warming the world while hunting to death and eliminating habitat through farming.

Heintzman is a frustrated idealist, entranced by *Jurassic Park* as a child but dismissive of any chance that dinosaur DNA could ever be found or made anew. Even under perfect conditions, it is unlikely that molecules of DNA could last more than a few hundred thousand years. But others, like the paleontologist Jack Horner, disagree, arguing that dinos can be brought back by reverse engineering from latter-day descendants like the chicken. I always knew there was something malevolent in the affectless eyes of the rooster.

Instead of reverse engineering a dinosaur from its much-removed descendant the chicken, why not focus on the birds themselves? The chicken is the best-studied avian species, closely followed by the nearly extinct houbara, a desert bustard with a flamboyant habit of raising brilliant black and white feathers when hunting a mate. The plumage makes the bird beloved of sheikhs with lots of oil money to fund research. Trailing behind is the humble passenger pigeon, a far more likely candidate for de-extinction than any dino. Through genetics, the idea is to get the physiology, the gestation, the skeleton, the color, the very shape of the passenger pigeon to exist anew.

But is that a passenger pigeon? Perhaps even more important, what makes for a species that distinguishes a passenger pigeon from its relatives the banded pigeon or the all too abundant rock dove, *Columbia livia*? The new bird will have the immune system of a band-tailed pigeon and the color of a passenger pigeon. So what is it? "There is no such thing as purity," Novak reminds me, while fiddling with a plastic Transformer robot he keeps on his desk. Messy mixing may be the fate of all life in the Anthropocene.

What Novak is really making is a chimera—a creature of Greek myth, late of what we now call Turkey though back then it was the kingdom of Lycia. The beast made an appearance in *The Iliad* as a creature made by the gods, with the head of a lion, the body of a goat, and the tail of a snake, sometimes translated or described as a dragon. All three heads survived this kludge, the lion in front, the goat from the middle of the back, and the snake's head at the end of the tail. None of those animal heads could save her, however, when confronted by Bellerophon, the grandson of too clever Sisyphus, astride the winged horse Pegasus, according to my tattered but lovingly illustrated edition of *D'Aulaires' Book of Greek Myths.* Homer, who diverted from the saga of the Trojan War to tell the tale (omitting Pegasus), made sure to emphasize that the monstrous creature was "of divine stock, not of men." But modern-day chimeras are all too human.

At a brown table in a conference room overlooking the trunks of redwoods, I discuss such chimeras and hybrids with this reviewer's board of paleogenomicists. What is a species, anyway? Carl Linnaeus canonized the most accepted definition back in 1753: a group of organisms capable of breeding and producing offspring capable of breeding. Thus the horse and donkey are species whereas the mule is a hybrid. But hybrids that can breed, like grolar or pizzly bears newly wandering the near Arctic or coywolves prowling a largely carnivore-free Eastern Seaboard, argue that this biological concept of the species is dead. For good measure, old Carl made himself the type specimen for *Homo sapiens*, though it's now clear we were happy to interbreed with Neanderthals, Denisovans, and other hominids.

In the meantime, as Heintzman reminds me, there are all the smaller animals to catalog and study. Ancient insect DNA may be the next frontier, especially since DNA can be extracted from something as small (and easy to preserve) as a wing scale trapped in amber. What happened to these sensitive indicators of climate and the environment roughly 40,000 years ago? "Let's look at lemmings," he suggests. "They're good for testing the human-versus-climate thing." As would be finding some of the drowned human settlement sites along what were the coasts of North America when the continent was first peopled all those years ago. But that requires copious funding for expeditions. So people go on with the relatively cheap work of sequencing genomes while leaving the larger questions unanswered and unexplored. "We can sequence a genome in a week, but no one has the money to pay for digs for ten years," Stiller notes.

Far away, in Nottingham, the Anthropocene geologist Zalasiewicz's home and Robin Hood's former haunt, there sits a repository for rare animals. Not the animals themselves but their DNA, frozen to −196°C and slotted away. This Frozen Ark stores the genetic information from 48,000 individuals of 5,438 different species, and growing.

The world may need that ark and others like it, such as the one in San Diego. Freeze them all and let our descendants sort them out. As it is, more than 300 species are known to have gone extinct since people came on the scene, starting with the dodo and ranging, most recently, to the baiji. The debate over the woolly mammoth may rage on, but there is no doubt that human influence—in the baiji's case, pollution, overfishing, and transforming its river home by building dams—killed the dolphin, believed to have been descended from a drowned princess–turned–river goddess.

Wildness today requires a deliberate act of human will, either to refrain from filling in a marsh in pursuit of another neighborhood or to actively restore what is now missing. Under the rubric of rewilding or just plain old restoration, wolves have been reintroduced to the western United States, beavers have been brought back to Britain after a 500-year absence and, maybe, one day, mammoth will again roam the Siberian tundra after disappearing for 4,000 years.

This is not an idea everyone cheers. Many people are happy with animals only as long as the animals don't get in the way, serve a definitive purpose like lab rats, or function as entertainment, either cute animal videos on the Internet or nature porn, a twisted reflection of life outside human concerns that nonetheless at least can inspire some people to care. And then there are the ecologists who care deeply but note that nature in some of these depauperate places has adjusted to the absence of missing animals or plants, and reintroducing them may fulfill some dimly understood human need but does not constitute restoration. In fact, rewilding may wreak havoc on the world as it is today in ways that are unpredictable. The only thing scientists know for sure about biology and ecology is how little they know—and how little can be predicted. "An insect is far more complex than a star," the astronomer Martin Rees likes to say.

The human relationship with wild animals in the Anthropocene has taken a turn toward the bizarre, the baroque late phase of decadence. On the one side, ardent conservationists are laying down their lives (and paying others to do so as well) to protect the dwindling herds of forest elephants in central Africa. On the other, game wardens in Romania feed brown bears so that other people can come and shoot them. As journalist David Quammen writes in *Monster of God*: "The gamekeeper's role, vis-à-vis his resident bears, combines the services and attitudes of a nanny, a zoo attendant, a field naturalist, a sniper's spotter, and a pimp." Similar relationships exist from the Karoo of South Africa to the icy wastes of Alaska, where big-game hunters pay big bucks to hunt an animal that has been kept alive specifically for that purpose. Cecil the Lion died, provoking international outrage, so a dentist could have a trophy. Meanwhile, people starve in Zimbabwe.

Such passion for hunting is both why the passenger pigeon is now extinct and why it may not be the first bird to benefit from the tools of synthetic biology. The houbara and its resplendent plumage is the favored target of oil sheikhs. And the Saudis, who love the bustard, have the money to spend to save it from themselves thanks to the modern world's addiction to oil.

Or perhaps the heath hen may beat the passenger pigeon to revival, pulling on the deep pockets of the residents of Martha's Vineyard, where they finally went extinct. Then again, the pigeon business is worth hundreds of millions of dollars globally thanks to humans known as fanciers. If even a fraction of those who fancy pigeons can be engaged in bringing back the passenger variety with their discretionary dollars, "We're home free," as Long Now's Ryan Phelan puts it.

The human instinct to hunt, which helped bring down the mammoth and the giant kangaroo, may be the last thing saving big mammals now. Hunters are the original conservationists and remain responsible for the majority of wilderness and wildlife protection the world over, like Wetlands International protecting unloved marshes for their value to duck hunters. The American bison was saved from slaughter by a captive-breeding program at the Bronx Zoo so that it could be hunted again. The auctioning off of a license to hunt an endangered white rhino raised money to preserve his fellows (as well as the ire of strict preservationists). Similar instincts expressed as aristocratic privilege have preserved wilderness areas such as the forests of Romania that house Europe's largest population of bears thanks to the hunter's lust of mad dictator Nicolae Ceaușescu. The sacrifice of a stag or bear each fall to a hunter's weapon can preserve both the natural character of the land as well as all the other inhabitants of the forest, as I can attest from family trips to the Northeast Kingdom of Vermont for deer hunting, where the real trick is not shooting a deer but getting a carcass out of the woods without roads.

This kind of ambivalent relationship to wildlife is likely to be a hallmark of the Anthropocene. Some, like Novak, will be dedicated to preserving or bringing back certain animals or plants. Others will be equally dedicated to killing them. Some of the most enthusiastic speculation around the TEDx de-extinction event came from readers of *Field & Stream*, who debated online which animal would be best to bring back to be hunted anew. The writer plumped for the heath hen, but the nominations included everything from the woolly mammoth to the Irish elk. Just imagine a mammoth's head on your wall or an image of it huffing and puffing down the trail captured on camera. Human purposes, whether of care or carnivory, remain paramount.

Humans are the world's worst invasive species, spreading out of Africa like a weed and proliferating wherever we go. Perhaps people hate our fellow invasives so much because they highlight our own flaws, much as parents tend to loathe any sign of their own worst traits in their children. We spread new life around the planet—wallabies living in a suburban forest outside Paris and the parakeet flocks of Brooklyn. The animals we've chosen thrive—the global livestock herd is now 50 billion animals. And synthetic biology isn't just about repairing what has been torn in Earth's great biology, it's also about speeding the rate of creation—and directing it toward new ends.

The best news for the passenger pigeon is that its old habitat is growing back and growing in extent. The eastern woodlands have returned as farms fade away or move west to more fertile ground—in the past decade, 40,000 square kilometers of forest have regrown. That has brought back deer, coyotes, turkeys, raccoons, and even black bears in abundance.

There is a home for the passenger pigeon and one of its dance partners of old, the chestnut; although eliminated by an invasive fungus from China, it may beat the bird back to the forest thanks to some similar genetic engineering efforts.

In the 1800s, the chestnut was even more abundant than the passenger pigeon, making up an estimated one out of every four trees in the Great Eastern Forest of North America that then covered the lands between Quebec and Florida and west almost to the Mississippi River. That's at least 3 billion trees, or roughly one for every passenger pigeon. But imports of Asian chestnut trees brought along a fellow traveler—a fungus commonly known as chestnut blight. The American version of the tree had no defense. Within decades, the chestnut was gone, killed off by humans, just like the passenger pigeon.

The same genetic engineering that might bring back the passenger pigeon has been used to create an American chestnut tree with the genes from the Asian variety that enable the tree to survive the blight. All that stands between the new chestnut and its return to the regrowing forest is federal approval for the release of a genetically modified organism. "You and your children and your grandchildren need to help with this because it's going to take a long time to bring it back," says the forest biologist William Powell of the State University of New York's College of Environmental Science and Forestry, who is leading the effort.

Plants should come first, since the passenger pigeon might not survive in the wild without chestnuts to eat. Plants make up 99 percent of living matter on the planet and provide all the food with the subtle trick of turning sunlight to sustenance via photosynthesis.

The redwoods that make Santa Cruz's campus unique may even need help. Walking among them on the soft duff amid the curling fern fronds, it is impossible not to feel the pull of a druidic sacred grove—as well as big lumber. This is an entirely human-made felling, grottoes sculpted by the hands of people. But our genetic tools are, as of yet, ill suited to the complexity of plants. The genome of a tree can have 22 billion base pairs, or seven times as much genetic information as a person—and we barely even know ourselves genetically.

Even if the genetics work out someday, it will be very hard to reintroduce anything even resembling a passenger pigeon back into the wild, as the chestnut example proves. The Wildlife Conservation Society, heirs of

an effort that saved the American bison at the turn of the twentieth century, now has a confiscated flock of band-tailed pigeons and is working to
train them as a potential host for any resurrected passengers.

Fortunately, humans have a lot of experience doing weird stuff with
birds, like ultralights teaching cranes to migrate and puppets feeding baby
condors. Novak's plan is to dye his band-tailed pigeon parents, so that the
chick can see a cosmetic replica of itself. Actual passenger pigeons were
not good parents, however. "They abandon their young before the babies
can even fly," Novak notes. " 'Oh, you're fat enough. Goodbye.' " That's
not rare among animals, but it is rare among birds, and no self-respecting
band-tailed pigeon would ever do such a thing.

How do we make that adult flock so they migrate like they used to?
Training homing pigeons, also painted, to teach the baby passengers
where to take a ride to, from late autumn roosts in upstate New York
down to southern forests, suggests Novak. Maybe we'll even train the
passengers to go to different spots than the historical record suggests,
thanks to the climate change that took hold during their time away. Such
assisted migration may prove vital in any human bid to preserve many
species, not just resurrected ones.

Then there's the question of law. Is a resurrected passenger pigeon—
once there is even one—automatically put on the endangered species list?
Or is it a GMO and therefore faces regulation before it can be put into the
wild, like the American chestnut? As the title of a provocative paper put
it, the biggest challenge facing any resurrected species may be "How to
Permit Your Mammoth."

Novak spends time grinding bones into fine powder from the catalog
of Shapiro's specimens: woolly rhino, mammoth, horse, even the polar
bear and, of course, his beloved passenger pigeon, represented by little
vials containing toes. Any surface that touches the bone is first sprayed
with a mixture of bleach and ethanol (to kill off any promiscuous DNA
floating around in bacterial cells or on people's gloved fingers) and wiped
down with paper towel. He squirts a liquid enzyme into the powder of
a passenger pigeon toe bone, which rips apart the bone cells. A buffer
is added to extract the DNA while the sample roasts in a rotating oven
for twenty-four hours or so. The cooking frees up the DNA from the

proteins that bind it. What's left is a simple soup of molecular DNA—the precious code, if you have the tools to read it.

A binding molecule is added to the soup, which seeks out bits of DNA and grips them tightly. The soup goes through a molecular sieve, the proteins pass through, but the DNA is caught. Novak washes the filter and puts the harvested DNA into a collection tube with a little bit more chemical buffer that looks no different from water but frees the DNA from its binding. A drop of this liquid can contain hundreds of DNA sequences or fragments—all destined for the technology that makes modern DNA work possible: polymerase chain reaction, usually known simply as PCR.

This was the first Nobel Prize–worthy breakthrough, back in the early 1980s. Essentially, the DNA is alternately heated to 98°C and then cooled to 55° while an enzyme—the polymerase from which the technique derives its name—guided by a fragment of synthetic DNA that acts as a primer grabs onto the desired bit of DNA. The polymerase then builds an exact copy of the DNA, and each copy is in turn copied until thousands of copies have been made. The thousands of copies then make the DNA easy to study and manipulate (lots of spares), even if the DNA is 700,000 years old.

Novak's bones are a little more than a century old, kept in a big cardboard box from which he pulls little bags, like some strange drug trade for science. It's not just toe bones but also a humerus or two and other bigger bones, lifted from several institutions through patient detective work and requests. At the end of the PCR process, Novak might find that somewhere between 9 and 35 percent—so no more than a third—of the DNA is native to the passenger pigeon. That compares unfavorably with, say, the 80 percent extracted from mammoth bones.

Those involved in de-extinction tend to see their work as part of a tradition of humans as gardeners, only now we're gardeners of the whole planet picking and choosing which plants and animals get to survive and thrive. Even the mighty and ubiquitous microbes are shaped by our choice of antibiotics, though we can most likely never set the clock on life back to zero.

This does not mean that humans are in charge of evolution, or anything like that. "We can't say we're in control if we don't know what we're doing," Shapiro notes. But in the wider world, people and their choices certainly set the parameters under which evolution operates: climate, habitable zones, selection pressure, and the like.

To take a famous example, St. Patrick allegedly rid Ireland of snakes. What will we do with a whole planet? Now scientists weigh whether to use genetic engineering to rid the world of the mosquitoes that spread malaria and other deadly diseases. That would be a calculated extinction, purposeful—and the kind of extinction once reserved for viruses like smallpox, which live on only in their own frozen arks in the bowels of medical facilities.

Then there is the question that intrigues Shapiro: Why did the mammoth die out while the caribou, still hunted today, did not? And what's to stop the mammoth or any other resurrected species from dying out again?

The passenger pigeon may prove a particularly bad candidate for de-extinction. Chimeras are hard to make even in well-studied species, like cows, where research has been going on for decades thanks to commercial interest. Even chickens have resisted such techniques, perhaps a residue from the dinosaur lineage that resists any Jurassic Park. And if the birds migrate, as the passenger pigeons did from roost to roost, well . . . "If you only have one or two and they fly away, you're gonna cry," warned Susan Haig, an ornithologist for the U.S. Geological Survey, at the TEDx event. Novak agrees: "If you've only got ten passenger pigeons, you don't want falcons coming in and dwindling that down."

Some think that spending millions of dollars resurrecting a particular species of bird springs full-blown from a somewhat dystopian sci-fi novel. "How many animals have to die so that one hybrid can live?" asks Shapiro, and others.

Still, the passenger pigeon makes as good a candidate as any: The eggs are not too big and the birds require minimal parental care, even if they do migrate. Passenger pigeons also have a small clutch size and an unknown, and apparently delicate, way of mating. However, as the eminent bird biologist Mike McGrew notes, "We don't have the technology to bring extinct birds back to life. We can't clone birds."

The easiest species to de-extinct may be one of us: Neanderthals, Denisovans, maybe even Australopithecenes. "It's the most feasible, but the worst ethically," says Stiller with a shrug.

Meanwhile, the mountain gorilla teeters on the brink of joining the

Neanderthal in the long lineage of ape extinctions thanks to us. Will we save our fellow apes, or will we allow them to disappear? "Preventing extinction is a bargain compared to reversing extinction," as even the de-extinction cheerleader Brand notes. The Long Now Foundation has partnered with the U.S. Fish and Wildlife Service to see if sequencing black-footed ferrets can reintroduce some genetic diversity to an animal that was once down to just seven furry individuals in the 1980s. "All ferrets today are basically cousins," Phelan notes, and that's not good for long-term viability.

The tinkering could prove even more profound. Church has suggested adding the genes that protect certain bacteria against the damaging effects of radiation into the genomes of would-be astronauts, to enable long-distance space travel. Or perhaps Neanderthals, Denisovans, and other extinct hominids with DNA records could be resurrected using the genes of another of our close relatives, still living: the chimpanzee.

No place is changing faster than the Arctic and, funnily enough, that might be one argument for resurrecting the woolly mammoth. The mammoth may re-create the mammoth steppe in Siberia, helping to trap carbon underground, and there's even a Pleistocene Park there awaiting the mammoth's return. "A part of the boy in me wants to see these majestic creatures walk across permafrost," says Hendrik Poinar. "But the adult in me wonders whether we should."

The woolly mammoth was still alive in what is now Russia when the Egyptian pharaohs were building their pyramids and the first, and perhaps legendary, emperor crafted a unified China. By the time of the Romans a few thousand years later, the mammoth was extinct. Was it nomadic hunters like the Chukchi, who still live near Wrangell Island, or the simple fact of climate change?

If the passenger pigeons are brought back, what is to stop us from eating them to death again? In the Anthropocene, cuteness may prove a survival skill.

The flip side: What makes plants and animals go extinct when it isn't us? Climate change is a big player, at least according to the fossil record. Is the global warming we're unleashing going to unleash a fresh apocalypse on fragmented species? "Those species at risk, a lot of those, this doesn't mean extinction right away," says the ecologist Dov Sax of Brown University. "Many of those may save themselves via evolved change. Some of those might be saved by humans."

"Imagine three hundred years from now. CO_2 levels were lowered in the atmosphere and temperatures are returning to something more like today. That might be a short enough period that you could avoid most extinctions. A lot of those extinctions take centuries to play out," Sax adds. "All hope is not lost, that's for sure."

For those species not yet extinct but at risk, there's always assisted evolution: humans helping direct the process of natural selection by culling individuals with low heat tolerance, say, or breeding individuals from farther south with those farther north to promote beneficial gene flow. A lot of weird, wild, wacky stuff will be tried, as much as human ingenuity can devise.

Perhaps there are better candidates for de-extinction than the passenger pigeon: animals or plants that provide some unique benefit to the ecosystem, or are evolutionarily unique, like the marsupial wolf known as the thylacine. Americans may still not be ready to cope with flocks of billions of birds. The world is in a triage situation in the Anthropocene. Tough decisions have to be made about what species to save outright, what species to save for later by freezing genetic data that may come in handy, and what species to, with due respect, allow to perish, as has been the norm on this ever-changing planet for billions of years. After all, stasis would be the true death of nature.

There is no ideal candidate for de-extinction, as Shapiro says, just those that are slightly better or slightly worse because of various characteristics. When experimenting with an extinct species, at least the pressure is off and there is no urgency. The worst has already happened.

Experimenting with embryos and cells from an abundant species like the band-tailed pigeon also means less worry of wasted resources. "We're using embryos to produce cell cultures. We're using surrogate parent birds. We're going to be using animal resources that for a very endangered bird are best suited for just breeding birds," Novak notes. "If we fail, we learn things are valuable for conservation and the world isn't left with another extinct species. And if we succeed, the world gets a new organism and all that information at the same time."

Even if the passenger pigeon effort ultimately fails, it does not fail to inspire people to dream of a different world. A world that could be full of the possibility of wildlife again, wildlife that rouses people to care. "We get to watch the passenger pigeon rediscover itself in the New England

and Great Lakes forests of North America," Novak says, including the trees in our backyards.

What kind of menagerie do people want to have around us and with us? On the Santa Cruz campus, I see three young bucks with the felt still on their horns with a doe, all gathered under the trees near the tennis courts just downhill from the redwood grove and not too far from the highway where the cars glitter in the strong California sun like sunset on a lake. They seem like a nice addition, one that I would be loath to lose.

This generation of people could oversee the end of the elephant, just like those previous generations that Novak castigates for extincting the pigeons he loves. So why work on passenger pigeons rather than getting out in front of the bullets mowing down the elephants? The answer is as simple as it is irrational: obsession.

Born under the sign of Capricorn, Novak is as stubborn as a goat when it comes to bringing back the passenger pigeon, partnering with Shapiro to study the likely impact on the landscape, people and, most important, the bird of bringing the living wind back. "I'm the guy who will fight through failures to make it right," he says, and he has embarked on a master's program at UCSC to try to puzzle out the lived experience of his beloved bird: how the population of the passenger pigeon worked in the context of the forest, and how it might work in the new forest regrowing today.

It's an animal passion shared by his family, who helped finance the first genetic sequencing runs of the passenger pigeon. The bird could serve as a flying "beacon of hope" highlighting the switch from killing off species to bringing them back, an evolution in human behavior toward more responsibility. Even the International Union for Conservation of Nature—the Switzerland-based independent arbiter that keeps a list of endangered animals the world over—has formed a committee to determine an appropriate de-extinction policy, just as it has formed committees to cope with rewilding, habitat loss, and other novel issues facing wildlife. "We're not a meteorite, we're not continental drift," Novak notes. "We can do something about extinction."

Every species is the answer to a question we don't yet know enough to ask. By saving it, we are saving ourselves from our own mistakes. A wild species of rice from India has already saved global rice production from an insect pest. Otherwise, we risk turning the Anthropocene into the

Eremocene: the Age of Loneliness as humans wipe out animals, plants, insects, fungi, and microbes without much thought, an alternate name proposed by the eminent Harvard biologist E. O. Wilson.

But "we're never going to do it unless someone wants to," Novak muses, pushing a flop of dark hair back from his eyes. There are more and more of those someones in the world as the human population swells, even as more and more of us crowd into burgeoning cities, even more removed from the forest woodlands that the passenger pigeon called home than the urbanites who ate the bird to death in the nineteenth century. Novak's greatest worry is that he will become a simple hype machine rather than a serious scientist working toward his resurrection goal. "I think the human condition is about exploring and doing new things and not being satisfied with what is normal or accepted. And so, from a purely scientific wonder-and-awe perspective, I think we have to pursue these things." Bring back the dead, says the would-be Lazarus Man.

Chapter 5

The People's Epoch

Have you entered the springs of the sea? Or have you walked in search of the depths? Have the gates of death been revealed to you? Or have you seen the doors of the shadow of death? Have you comprehended the breadth of the earth? Tell Me, if you know all this. Where is the way to the dwelling of light? And darkness, where is its place, That you may take it to its territory, That you may know the paths to its home? Do you know it, because you were born then, Or because the number of your days is great? Have you entered the treasury of snow, Or have you seen the treasury of hail, Which I have reserved for the time of trouble, For the day of battle and war? By what way is light diffused, Or the east wind scattered over the earth? Who has divided a channel for the overflowing water, Or a path for the thunder-bolt, To cause it to rain on a land where there is no one, A wilderness in which there is no man; To satisfy the desolate waste, And cause to spring forth the growth of tender grass? Has the rain a father? Or who has begotten the drops of dew? From whose womb comes the ice? And the frost of heaven, who gives it birth? The waters harden like stone, And the surface of the deep is frozen. Can you bind the cluster of the Pleiades, Or loose the belt of Orion? Can you bring out Mazzaroth in its season? Or can you guide the Great Bear with its cubs? Do you know the ordinances of the heavens? Can you set their dominion over the earth? Can you lift up your voice to the clouds, That an abundance of water may cover you? Can you send out lightnings, that they may go, And say to you, "Here we are!"? Who has put wisdom in the mind? Or who has given understanding to the heart?

—JOB 38:16–36

Cun Yanfang, who goes by Angela among her English-speaking friends, used to spend her days collecting fuel wood—branches, sticks—from the local oaks. Though she has two brothers, sometimes she'd carry as much as 30 kilograms of wood on her back alone, all harvested as sustainably as possible. Care had to be taken to make sure not to kill the trees so that more fuel could grow back. And her brothers did help her plant more trees near her mother's farm in the foothills of the Tibetan Plateau.

The Cun family escaped China's one-child policy as members of the Naxi minority group, a people of 300,000 souls or so once described as the forgotten kingdom of what is now Yunnan Province. By the time Angela Cun was ten, the sound of axes filled the forest. The old town of Lijiang built by the Naxi people had become a tourist attraction for millions of Han Chinese, as had the snowcapped Jade Dragon Mountain beyond— and wood was needed more than ever before. All that tree cutting sped the river's work of eating away at the feet of the mountains, thrust up by the geologic collision of India and Asia as much as 55 million years ago. Even the trees near the spring from which her village drank began to be felled by desperate locals in need of fuel. Cun's young mind could see that this meant the old ways—harvesting wood from carefully chosen lots that rotated between families—were finished. No care for the next generation could withstand the crisis of the day. A flood and fouled drinking water were inevitable. "This touched my heart," she recalls.

The felling of the forests and the muddying of the waters overturned centuries of local tradition, which forbade dirtying the river, even with spit. The Naxi have farmed the river valleys of the Tibetan foothills since at least the Tang Dynasty, more than a thousand years ago, living on the floodplains of the Jinsha and Nu Rivers. "It is our taboo," Cun says. "I was upset. We really enjoyed our childhood in nature."

The Naxi way of life is slowly dying, as is the language, drowning under the rise of Mandarin Chinese taught at school and spoken in the streets. Even Cun's own nephew cannot speak the Naxi tongue any longer, bringing to a close what has been a carefully shepherded separate culture that has endured among the Han for at least centuries. "It's hard for our small culture to resist," Cun notes.

As a result, years later, after becoming the first woman in her family to study at university, Cun has come back to these valleys in an attempt to save something of the woods in which she used to lie on her back for hours as a child and just listen to the wind sighing in the trees. "When I am away, I miss my home," she says. "My friends say, 'You are alone. What's to miss?' But I want to go back. I miss the nature."

Cun became a tour guide, a safe and lucrative career. She even married. But she was still not happy. So she decided to become an environmentalist, much to her family's chagrin although they supported this controversial choice that might cause her to run afoul of the authorities. "I have always loved nature since I was little," she says. "We want to have a better life and still have beautiful nature." She even divorced her husband, who could not understand her passion for her work, too traditional and stuck in his ways.

Women in China, as Mao Zedong famously said, hold up half the sky and theoretically are on an equal footing with men in this would-be egalitarian society. Yet the mightiest members of the Communist Party that Mao guided to power are men. For a group that holds up half the sky, there is a noticeable lack of women in leadership roles, though there is usually at least one in a minor role at every ministry. In fact, China's one-child policy, paired with cheap ultrasounds to determine a baby's gender, has resulted in nearly 120 boys born for every 100 girls in China, compared to roughly 104 boys born for every 100 girls in the United States. In other words, there are fewer and fewer women to hold up their half of the sky, a daughter deficit.

Yunnan Province is the wild west of China, a land of mountains and jungles, and a gateway to both the drug and wildlife traffickers of Myanmar and the sensitive nature of Tibet. The great rivers of Asia start here, tumbling down from the highlands, waterways that will become the Yangtze, Mekong, and Salween are here known as the Jinsha (Golden Sands), Lancang (Mother of Water), and Nu (Angry River). The tall mountains still look exactly like those in the ancient Chinese scroll paintings that celebrated this wild nature despite the long history of people here, preserved by being too steep to farm or log. This is a true anthrome, shaped fundamentally by the actions and choices of people for hundreds if not thousands of years as the wonderful waterworks of Lijiang attest.

Liming lies deep in the mountains surrounding Lijiang, a tiny village ensconced within one of the last wild places in all of China, where the famous golden monkey still roams free, for the moment.

Like most frontiers, the focus for people in Yunnan is resource extraction: logging has denuded the sheer hillsides, replacing deep, dark forests with plantation pines as a result of the transfer of ownership from local villages and their rotation method of tree cutting to state control and commercial logging. The red earth, exposed, is a raw wound in the sides of the hills and mountains, as a logging path provides the route down for valuable trees. The deforestation got so bad that the state banned commercial logging, yet satellite observations show that the old trees are still disappearing from supposedly sacred and state-protected remnant forests, particularly those closest to sites for ecotourism.

A growing population doesn't help. The villages of the valleys, like Liming, used to be dark in the daytime as the surrounding trees and mountains blocked out the sun's light. Now those tall trees are gone, replaced by spindly pines, and the focus is on bringing a road and electricity to every village and a television to every household. Yet rock slides from dissolving mountains no longer held in place by trees routinely block these new roads.

Still, the people need wood to fuel cooking and heating in the cool valleys. Traditional stoves hulk in the village kitchens, belching soot and wasting wood. Better cookstoves are one potential solution, but traditional sensibilities must be respected. "Some cultures like to see the flame, so we cannot close it," Cun explains, and a closed stove is a more efficient stove. The Naxi use even more wood to cook the slop for their pigs. "We think the meat is more tasty," she adds. Farmers explain the practice by noting that people like their food cooked too.

At the same time, natural products from the dwindling forest, like mushrooms, provide the bulk of the livelihoods of local peoples like the Naxi. The desire to save the forest is still there, which is why the local Liguang Xinwen school has "Earth Helpers."

The day I visit the school with Cun, a light rain falls, driving us inside, where a small censerlike flame bowl filled with firewood heats a spare room. Students work next to the smoke wearing gloves with the fingers cut off, studying and furiously making marks on papers. In the corner lies a huge sack of mung beans, food for weeks to come. The students' English is flawless, and the Han teachers practice another local language, Lisu.

In this tiny school, the principal started the Earth Helpers club back in 2004. Its mandate focused on environmental protection, educating students who in turn talk to their parents. What was once just forty interested young minds has now swelled to include 60 percent of the children here, teaching them about endangered species such as the local Asian black bear and Chinese yew tree as well as keeping the school, village, and even individual homes free from trash. "Every student has applied,"

Principal Dong tells me through Cun's translation, giggling a bit at his success. "But the teachers have criteria, not all can join."

I watch the students recite the Earth Helpers pledge with a raised right fist. Then I stroll through the latest examples of their artwork, including a prizewinning painting of a logging truck driving away, followed by a flood. Some of the students fantasize about being superheroes and stopping environmental damage. One offers this poem in English, complete with illustrations: "More and Less."

> If more smoke in the sky, birds will be less
> More boats in the sea, fish will be less
> More people in the mountains, animals will be less
> More cars in the forest, trees will be less

The Earth Helpers have rules at once obvious and complex: Don't kill the golden monkey. Don't use paper cards for the annual Spring Festival. Clean the campus every day. Don't pour rubbish in the stream (a message that does not seem to have reached all the parents in Liming, given the garbage on the banks of the stream near town, a Naxi no-no). Learn how logging leads to floods: Some people cut the trees at the front of the river, and when the rains come there is a landslide followed by the flood. So, along the local Jialing River, don't cut trees. And use less fuel wood.

The students even convinced the teachers to quit smoking—for their health—two weeks before, though it remains to be seen how long that will last.

I was born in 1972, the year that global population growth rates peaked and began to decline, a trend that, despite what you may have heard or read, continues today. And yet a population explosion is still working its way across the globe. Humanity will likely peak at 10 billion people or so this century, give or take a billion, absent a major change in breeding.

This big number is a challenge, no doubt. The world adds more than 150 people a minute, and nearly 100 million hungry mouths join the human cohort each year. Today, at least 1 billion men, women, and children are chronically malnourished or outright starving. At 9 billion total by 2050, simply to keep the number of starving at 1 billion people could require clearing 900 million more hectares of land to grow more food. That's the equivalent of nearly another entire United States covered in new farms.

But with 4.3 billion hectares already in farming today, there just isn't that much room to grow, even if people were willing to sacrifice the rest of the rain forests and much else. People already employ—mostly for agriculture—40 percent of the land surface on the planet. The largest land ecosystem is an anthrome: pasture, thanks to our taste for meat.

Fortunately, the world already grows enough food for everyone to consume more than 2,000 calories per day, except the food is not distributed that way. Instead, the rich enjoy more than 4,000 calories per day while the poorest starve. A dog in the United States or Europe may well eat better than a person in Bangladesh or Gambia. The richest 500 million people on Earth spew half of the greenhouse gases, changing the climate with their cars and electricity use while the poorest 3 billion contribute just 7 percent, mostly through those cooking fires and the tree cutting that fuels them.

A child lucky enough to be born in the United States today, whether a girl or a boy, will add an estimated 10,000 metric tons of carbon dioxide to the atmosphere in his or her lifetime. Each American, some more than others, has an outsized impact on the world. An average American requires 22,000 kilograms of metals and minerals wrested from the ground each year, most of it elements used unthinkingly, like stone, sand, gravel, and cement. There's also the salt and phosphate needed to grow the crops we consume to live. And then there's the roughly 4,000 kilograms each of the fossil fuels: coal, natural gas, and oil.

The average American uses 90 kilograms of stuff each day, day in and day out. We consume 25 percent of the world's energy despite being 5 percent of the world's population. We lust for the latest gadget, which hides away minerals wrested from beneath the Congo, among other places, deep in its innards.

That means forgoing a baby is the single most impactful decision a couple will make. One fewer American child reduces greenhouse gas pollution twenty times more than driving a Prius rather than a pickup truck, using energy-sipping appliances, or any other lifestyle choice available—combined.

The year 1987 appears to have been peak baby—87 million people born. Today, only 78 million new babies join us each year. Human numbers are dropping voluntarily, and not due to challenges from disease or natural disaster as in the past. As recently as the 1960s, the average family around the world had 5 children; today the average is 2.6. As the journalist Fred Pearce writes in *The Coming Population Crash*: "The 'population bomb' is being defused. By women. Because they want to."

Population rises fastest in places where women are not in charge of their own bodies and where they cannot own land, among other age-old repressions. Family planning works, whether in the United States, Iran, or Thailand, crossing cultures. Empowered women are the reason that the human population growth rate peaked decades ago at just over 2 percent and is now a little more than 1 percent. This economic and reproductive empowerment must continue to ensure an enduring Anthropocene.

How many people is the right number? No one can say. Back in 1679, the pioneering microbiologist Anton van Leeuwenhoek calculated that the planet could hold 13.4 billion people, based on the population density of his native Holland and its size relative to the rest of the globe. Modern guesses are no better nor more scientific, ranging from 1 billion to 1 trillion. Worrying about population growth is an old game. As early as 1600 B.C., Babylonians feared the world was already too populous, and that's when the planet held maybe 50 million people, or roughly the population of the megacities Tokyo and Mexico City today.

What matters most is not total numbers but, as the American baby proves, the relative impact of each individual. Only a small percentage of that big 7 billion number contributes most to the problem of a disappearing world. And consumption is the real challenge, people appropriating every hectare of land, every cubic meter of the sea, and every molecule of the air for our own comfort and support. That leaves nothing for the inedible plant or the wild animal. We are allowing the workings of the world without us to be lost in a world of homogeneity, a subset of plants and animals that we tolerate and that thrive alongside us or that serve us. It is a new Pangaea, or union of continents, convened even as the continents remain separate, due solely to the hypermobility of one species.

I sometimes see rats on my commute to work, the outcasts searching for food down on the subway tracks. I always see pigeons strutting the city streets. These are just two of the animals that, aided by humans, are undergoing their own great acceleration. On the other side of the ledger, my morning cup of coffee aids the die-off of the black-handed spider monkey, a cute New World simian that thrives in the canopy of Central American forests in tall trees. That's pretty much the opposite of the kinds of trees that still exist where the monkey lives, because for decades farmers have been cutting down those tall trees to clear land for crops, among them coffee beans for the international trade that brings the habit-forming drug to my door. And it is our choices that will save or condemn the lively monkey, and if it is to be extinction, the taste of coffee may become too bitter to bear.

My vanity and social obligations suggest I wash my hair and brush my teeth, but the fungicides and plastic microbeads in those products have their own far-reaching journeys to undertake and damage to do. I ride to work on a screeching train, shuttling through tunnels carved underneath Long Island and Manhattan, even under the East River itself. The electricity that propels those subway cars comes from atoms split a few tens of kilometers north at the Indian Point Nuclear Power Plant (now renamed the Indian Point Energy Center) and molecules of methane burned at power plants scattered throughout the city. I am entertained en route by a tiny smartphone that has more computing power than was available to the engineers who put the first men on the moon. I use it mostly for apps and reading. My phone and laptop contain enough precious gems and metals to put medieval popes to shame.

I work in a building that struggles to control its internal temperature, fighting off the summer's heat with too much cooling and the winter's chill with a blast of warmth. I sit on a chair molded from plastic and fabrics created from oil, atop fibers in a carpet crafted from oil, and type on a keyboard made mostly from oil. For lunch, I head out to a local restaurant and enjoy food prepared from ingredients carted in from the countryside by ubiquitous small trucks burning diesel and belching sooty exhaust responsible for higher rates of asthma near New York's highways. I sip another cup of coffee in the afternoon that kills more monkeys just to boost my brain a little, so I can write like this. I use the bathroom, flushing away potable water that billions of people lack and wipe my butt and hands with the pulpy remains of tree plantations that have replaced native forests.

At the end of the day, I eat more food trucked in from afar, possibly even delivered by car as well from some restaurant nearby, maybe even pizza from a coal-fired oven to maximize my greenhouse gas contribution. The flowers I give my wife may say nothing so much as: Look at me, I'm happy to kill for love. Then I relax with my family, enjoying television or a movie displayed on a screen made flat by the ingenious use of yet more precious metals and minerals. Those movies are environmental extravaganzas, lighting the night, building fake cities, employing more computing power than once tracked the possibility of global thermonuclear war to conjure up living nightmares.

Or perhaps I find myself in hotel world, that safe, anesthetized cocoon replicated across the globe with synthetic fossil-fuel-based fabrics, electric light, and charging outlets. If I do, I most likely got there by airplane, spewing CO_2 high in the atmosphere where it can trap heat for millennia along with the water vapor that forms contrails, scarring the sky, at least for a time.

I take off my cotton clothes and slip beneath cotton sheets, the natural fabric employed in vast quantities for its softness and durability but available only thanks to a copious quantity of deadly pesticides made from fossil fuels that can poison the land. Then there's all that water used to grow puffball blooms in the harsh desert of Arizona, so much so that the Colorado River no longer reaches the sea in Mexico, an entire paradise of marsh and wildlife drunk up to provide cities in the desert with flush toilets and bountiful crops from California to New Mexico. I sleep well because all this can be ignored, not exactly safely but for longer than my forty-odd years on the planet. Modern life is a series of abstractions that insulate the lucky among us from reality, whether it is where meat comes from or who pays the cost for our gadget lust. As the Israeli historian Yuval Noah Harari likes to say: "Consumerism is the first religion in history whose followers actually do what they are asked to do."

And this present I take for granted is but a bright, gleaming hope in some still distant future for too many.

The world is getting better: Only 46 out of 1,000 children born globally this year will die before the age of five from tuberculosis, malaria, malnutrition, or a host of other diseases. That's down from 90 children born in 1980. This fall in infant mortality has also helped keep fertility in check: More babies survive, so families have fewer children. But there are still more than 3.5 million dead babies every year from ailments that have been eliminated almost entirely wherever there is access to modern energy, medicines, and care. And then there are all the adults who die too young.

Disease remains the big enemy of people. The Spanish flu killed 20 million in the early decades of the twentieth century. Smallpox and a bevy of other diseases depopulated the Americas just in time for the Spanish, English, and French to run roughshod over the remaining inhabitants and lay claim to a new world. The early accounts of vast cities were so out of whack with the empty landscapes found by later explorers that they were dismissed as myth. But the mounds at Cahokia tell a different tale. Now it is Ebola, along with the killers that have been with us from the beginning, like cancer.

A Zulu crowd will ululate whenever given the chance. This time it was to welcome Jacob Zuma, president of the Republic of South Africa, upon his return to KwaZulu-Natal, his native province. He had come to reassure the excitedly chattering men and women of iLembe District of his ongoing commitment to bring them, his people, and the rest of the nation more energy. Roughly a third of the people in the richest nation in southern Africa lack access to electricity. He even brought gifts, in the form of sun-powered water heaters, LED lights, and cookstoves that trapped smoke rather than spewing it into the home. That smoke is why poorer Zulu suffer from high rates of tuberculosis and asthma.

The gifts actually came from various United Nations organizations, which helped install solar hot-water heaters on five hundred homes and gave to two hundred homes the LEDs powered by specially engineered wafers of silicon that turn sunshine into electricity. "Let us change, instead of burning, making more smoke," Zuma said. And then, as if staged to illustrate the urgency of the need, the tent on the soccer field in KwaDukuza went dark.

The diesel generators powering the lights on this rainy day in South Africa had stopped.

The world today is in the midst of a neo-fossil age, an era of burning ever more. We don't even burn fossil fuels particularly well, transforming perhaps 40 percent of the potential heat in coal into electricity. Those black rocks are fossilized sunshine, remnants of the first woody plants. Coal's cousins oil and gas come from similar sources—the corpses of ancient life that trapped the energy of sunlight in innumerable cells. The next trick we need to invent is a way to cut out this residue of abundant life from the geologic time series and go direct to that sunshine, that convenient fusion reactor at a comfortable remove streaming forth an abundance of energy. Winds stirred up by the sun's unequal heating of various parts of Earth's surface can be harvested, or even Earth's heat itself can be tapped for steam.

Zuma's rally was held back in the "year of sustainable energy for all," 2012, if you missed it. There is a new goal being worked on by aid and development types, an effort to create global access to "modern energy services." That's for the more than 1 billion people without electricity, who might enjoy refrigeration and clean light for the first time. That's also for the 3 billion people who still cook by burning wood, charcoal, or even dung to get access to a cleaner flame, or at least a better stove that can trap the lung-choking smoke. And all by 2030, if UN secretary-general Ban Ki-moon gets his way.

This tall, thin broomstick of a bureaucrat who has run the UN since 2007 grew up doing his homework by the dim light of a smoky kerosene lamp. Kerosene comes from oil and powers jet engines. It stinks and is dangerous, its only attractions being its portability, energy density, and the wan light it casts. Only at exam time did he get permission to upgrade to the slightly more bright light of an expensive candle. He was born in the pre-electric darkness of South Korea and was a freshman in college before he had reliable access to electric light. Meanwhile, his children take that same electric light—and electricity in all its glory—for granted. South Korea has become one of the most advanced nations in the world in the space of a single generation. It's a feat he would like to see repeated everywhere in the world.

All the stench and most of the danger of lighting is transformed by electricity. A lightbulb delivers a clean, pleasing light and the only danger is of a shock, or perhaps the glass of a burst bulb. It's cheap too; an incandescent bulb can cost less than a dollar, discounting the expense of the electricity-generating system required to power it. "We need to turn on the lights for all households," Ban tells anyone who will listen.

In Gambia, even hospitals don't have electric lighting or even electricity after certain hours. That means no power for medical devices that can save a life, whether an ultrasound machine to monitor a baby's health or lights to guide a surgeon's hands during a cesarean section to save both mother and child. Hospitals here usually lack incubators, which could save the life of a baby born too small or too soon. Why waste the money on a life-saving machine without the power to run it?

The list of medical wants electricity enables is long, starting with refrigeration for lifesaving vaccines and medicine. The lack of electricity kills; life expectancy in Gambia is sixty-one years.

At the same time, how the world gets such services provided by modern energy will play an outsized role in what kind of a planet we all live on. If the poor in India light the night by burning coal in power plants, as China, the United States, and Europe have done before it, then the climate will change even more significantly. This is India's true exceptionalism: The soon-to-be most populous nation will have to chart a new path for development. And if South Africa continues to mine and produce the rich seams of brown and black coal laid down in Carboniferous times, then smog will continue to plague its countryside and its people. If we all continue our obsession with hurtling about in several-ton vehicles made of metal, then the sacrifice of more deltas like the Niger in Nigeria and the Mississippi in Louisiana will be required.

In other words, if people all over the world adopt the American dream without changing the source of its power—coal, oil, and natural gas— then global civilization may face a prolonged nightmare. This problem is only going to get harder as the population of the whole continent of Africa may swell beyond 4 billion this century, all of whom deserve modern energy.

Fortunately, there is an alternative.

Kenya—where less than one in five households enjoys electricity—is attempting to become a green superpower. The would-be greenest nation on Earth is powering up in a new way, with a combination of harvesting Earth's heat, conveniently exposed by the Rift Valley tearing the continent apart to form a future sea, as well as wind farms and solar panels paired with batteries to store electricity for nighttime use in remote villages. Even nuclear reactors may be built someday in the not too distant future to replace any remnants of coal. That means reliable electricity in the grids of the new, big megacities like Nairobi and the minigrids of smaller towns, enough to at least charge a cell phone and light a bulb in more remote villages, with the hope for more in the future.

But the real locus of energy poverty is even more basic: cooking. Charcoal remains the fuel of choice for billions, and that not only means fewer forests, more soil erosion, and more missing animals and plants, it also is bad for people's health. People in the forest are the ones who bring

Ebola out of the forest, and burning charcoal in poorly ventilated homes cuts decades—yes, decades—off the lives of children, women, and men.

A full legion of smokeless cookstoves has attempted to solve this problem over the decades, so far without success. I saw an example from Dutch consumer goods giant Philips—a gleaming metal pail atop a little black saucer, hiding a roaring fire. Tongues of flame lick the top of the pail, where fuel is added or a pot rests, but no smoke at all wafts out. Charcoal, twigs, dung—this stove can burn almost anything. And that's by turning it into a gas first.

Like many poorer countries, much of South Africa smells of smoke. The open flame of indoor cooking adds the taste of fire to home-cooked meals for billions of people. At the same time, more than 2.5 million people die prematurely from the soot that accompanies those flames. Many others suffer from chronic lung ailments. And the soot that isn't inhaled wafts into the atmosphere and helps to cook the planet. It's just one big fire that keeps growing and growing and growing, now raging in China and, soon, India.

Our relationship with fire is primordial. In fact, some anthropologists argue that fire shaped us as much as we shaped fire: cooked food unleashed bigger brains, as did the stories told around flickering campfires. Now we use fire to reshape the world, whether ancient Australians burning their way to a different flora and fauna or modern Americans changing the mix of gases in the atmosphere to power vast cities and launch metal rods into the space we only recently discovered beyond the boundaries of the atmosphere. The myth of Prometheus continues to haunt humanity, fire as the first and still preeminent technology, as does the story of Faust for the pact we have made with fossil fuels to build and sustain modern civilization.

Part of the solution has been obvious for years: cleaner cooking. That could be the kind of advanced cookstoves offered by Philips, among others. Or it could be natural gas, as is common in many richer countries, or perhaps even gas made from garbage and sewage digested by microbes in a backyard pit. But there is a problem, a mismatch between the skill and time needed to tend these technology options and the daily lives of people who need them. The sun can be used for cooking, but such stoves obviously don't work after sunset, when the evening meal is prepared in many cultures. And the expense of a cleaner cookstove or a methane-making artificial stomach in the backyard completely surpasses a household's ability to pay, let alone maintain, if it is lucky enough to receive it.

As for Philips, the multinational's clean cookstove relies on modern energy to work—electricity provided by a battery to a fan. That fan blows air into the metal pail, which makes the flame burn hotter and more completely, eliminating smoke. All it requires is a relatively expensive rechargeable battery and an electricity source with which to recharge it. That may make it unworkable for the people it is intended to help.

The people of iLembe have access to intermittent power—after all, South Africa is one of the brightest oases of light in satellite pictures of the dark continent at night. That darkness is a stark contrast with the tendrils of light that create glowing webs connecting cities and communities in North America, Europe and, increasingly, Asia. Still, roughly 30 percent of households lack access to electricity, even in South Africa. The rest of Africa south of the Sahara is worse: 70 percent of households in this vast region lack both electricity and fuels other than wood and dung for cooking fires. What light there is in much of sub-Saharan Africa glows from cheap lamps that burn kerosene or paraffin.

There are alternatives here too. Some lucky individuals and communities can boast a bespoke lighting system, connecting solar panels on a roof to batteries inside that can store the energy chemically. At night, the batteries release that energy to power light-emitting diodes or LEDs, a high-tech source of pinpoint light. Families can supplement their income with new businesses like cell phone charging. But the first hurdle remains simple and obvious: cost. And then there are secondary hurdles like corruption.

For the lucky in South Africa, there is the grid. The nearby port city of Durban hosts a utility that boasts a wind turbine spinning briskly in the sea breeze blowing over its headquarters. But the reality of grid power in South Africa is abundant local coal, which the country also turns into diesel fuel for trucks and jet fuel for airplanes in vast industrial refineries that belch pollution constantly. The country burns coal and helps transform the global climate so that its wealthier citizens can have light at night—and electrified fences to keep their neighbors out.

At Zuma's rally, a tall, smiling puppet woman wears a witch's pointy shoes, striped stockings, and a red-brimmed hat with smoke coming out of the top. Her name is Mrs. Coal. She calls on an equally tall friend—a floppy, green walking wind turbine puppet known as Mr. Wind—to help her provide power. But the wind doesn't always blow, and if even rich countries cannot clean up, what hope is there for the poor?

Hope sprang anew in March 2010. This hope came just after the gloomy failure to reach a global consensus to combat climate change at Copenhagen at the end of 2009. A fledgling federal agency held its first summit in the cavernous ballrooms of the Gaylord Convention Center ensconced within the new urbanist nightmare known as National Harbor, Maryland.

The convention hall filled with the excited buzz of wild-eyed people whose ideas mostly passed muster with the laws of thermodynamics. Unkempt hair, bad sports coats, and excited stammers belied too many good ideas to fund, in the words of founding director Arun Majumdar, a onetime mechanical engineer and aficionado of materials that turn heat into electric current turned functionary with the cheerful but aggressive demeanor of Captain Kirk from *Star Trek*. This was during the heart of Silicon Valley's romance with what the tech bros dubbed "cleantech" and before the souring caused by the slow pace of an energy transition or a learning curve compared to the rapid growth available in apps or chips. The penthouse party was indeed in a penthouse and cleantech attracted gamblers like Neil Renninger, a onetime member of MIT's card-counting blackjack hit squad and until recently chief technology officer for synthetic biology pioneer Amyris. Why? Perhaps because energy is the only market where revenues are measured in trillions.

The venture capitalist Vinod Khosla, one of the founding fathers of the tech boom at Sun Microsystems, stood onstage as an expert to tell eager attendees, "Don't listen to experts." Experts are almost always wrong, he averred, and listed failures of expertise: dismissing the telephone, flight, home computers; an abysmal record for predicting oil prices or mobile phone growth. "Experts are as good as dart-throwing monkeys," he informed the worshipful crowd, perhaps insulting monkeys.

So expect a black swan, Khosla predicted—a rare innovation that goes on to have extreme impact and seems inevitable when looking back from the future. Perhaps, I thought, along with many others, no doubt, such a black swan hid as an ugly duckling at this very event, like the entirely new word "electrofuels" invented to describe a novel effort to use microbes fed on electricity or hydrogen to make fuel directly from excess CO_2 in the air. Khosla advised "more shots on goal," mixing his metaphors, to ensure such a black swan appeared. "Turn CO_2 from a problem to a feedstock? That's a black swan."

If experts are always wrong, then Khosla proved himself an expert in cleantech in those years, investing in the biofuel companies Kior and Range Fuels, which would ultimately fail. He was right about at least one thing, however: The more shots, the more likely to score, whether that goal is a better biofuel or a cheaper way to move electricity around the grid. Besides, Khosla's failures still lay in the future, and in early March 2010, hope wafted along in the recirculated hotel air, helped along by the scent of federal stimulus moneys, which brought a gleam to the eye of everyone from General Electric CEO Jeffrey Immelt to the proponents of a new kind of nuclear reactor devised at Sandia National Laboratories in New Mexico that works great on paper but doesn't yet exist anywhere in the real world. The stimulus may not have built a Hoover Dam, but it did leave behind this little agency that would if it could usher in an energy revolution, and all for the low, low price of $300 million or so in tax moneys per year. That little agency's name is a mouthful: the Advanced Research Projects Agency-Energy, more commonly known as ARPA-E to its fans and fundees.

At the agency's inaugural summit, I wandered up and down the rows of booths in search of this black swan. The booths showcased 120 different innovations, some of which spilled out of the ballroom and into the hallway, like the flying wind turbine. Although this electricity-generating kite was too big for a booth, it, and all the other innovations, seemed too small somehow to take on the world's largest machines—the electric grid or the gargantuan human enterprise known as the oil and gas industry— or the world's largest problems like energy poverty.

ARPA-E was an offer Majumdar just couldn't refuse. Leave your wife and children in Berkeley, come to a Washington, D.C., gripped by bipartisan dysfunction, and start a new federal agency with a limited budget. Oh, and take a pay cut. What fortysomething emigrant from India wouldn't jump at the chance to use DARPA as a model for energy research with a fraction of the funds?

DARPA, the Defense Advanced Research Projects Agency, started its run in 1958, created to serve the needs of the various parts of the Department of Defense for continually evolving technology. As a result of almost sixty years of DARPA's work, the world now enjoys the Internet and night vision, as well as more lethal assault rifles and self-guided bombs.

ARPA-E's mandate boiled down to a simple concept: ensure that the American children of today did not face diminished prospects compared to their parents or grandparents. An august committee that originally put forward the idea for ARPA-E specifically cited energy as the key to that, what with the challenges to a good life posed by climate change and economic competition. After all, ARPA-E is not alone in the quest for a breakthrough in clean energy, whether the competition is China's long-standing 863 Program for advanced energy technologies or Denmark's national effort to go carbon neutral. The list of potential energy research for ARPA-E is long: "ideas that could make solar and wind power economical; develop more efficient fuel cells; exploit energy from tar sands, oil shale, and gas hydrates; minimize the environmental consequences of fossil-fuel use; find safe, affordable ways to dispose of nuclear waste; devise workable methods to generate power from fusion; improve our aging energy-distribution infrastructure; and devise safe methods for hydrogen storage."

ARPA-E has worked on many of these. And what works for American children could yet work to save non-American children and adults. Majumdar and his boss Steven Chu's plan was to create a government agency that did not fear risk, a bureaucracy without bureaucrats to create that future. "We wanted to be measured in our craziness," Majumdar told me a few years later. "It is early and you want to create a reputation of solid foundations in technology which are risky but not wacky."

In addition to being the Indian avatar of Captain Kirk, Majumdar shares that character's penchant for boldly going where no one has gone before as well as directly challenging all those around him. He calls that "constructive confrontation," and it did help to winnow out the proposals and projects that violated the laws of thermodynamics. Constructive confrontation in the form of a barrage of questions also kept those ideas that were merely outrageous moving forward. Just one in one hundred proposals received funding in the initial round of ARPA-E projects, including $7 million to develop a battery that stored energy in liquid metals. Progress would be measured in CO_2 molecules not in the atmosphere, lungs not choked by smoke, landscapes not dug up for coal or pierced for oil, though it might take fifteen or even twenty years for that not to show up.

This is an economic and geopolitical challenge as much as a scientific one. Energy is tied to the fate of modern empires: Britain ruled through coal while the United States and Russia benefited from oil and nuclear. The people who invent (or control) the abundant energy source of tomorrow could well prove imperial once again, whether it's solar power from China or fusion from the European Union. Even Google, the twenty-first-century technology titan, has gotten in on this energy race with its RE<C program, a bid to make renewable energy from wind turbines or solar panels cheaper than coal (that ultimately failed). Their first try involved computer algorithms to better control lighter mirrors that tracked the sun to focus the heat in its light to produce power.

The star of that inaugural ARPA-E show was not the wild-eyed investor Khosla or even the charismatic Majumdar, but rather Chu, the most overqualified man ever to have served as U.S. secretary of energy. A self-proclaimed "Jackie Robinson of nerds," and recipient of the Nobel Prize in physics in 1997, he is the first scientist to serve at the cabinet level. There is something of the owl in Chu's heart-shaped face, both the proverbial wisdom and the predatory instincts, ready to strike the intellectually unprepared. He looks casual with his hands in the pockets of the dark blue power suits he favored while in Washington, but he attacks with questions. If energy could be personified, it would be Chu, dashing from one global event to another all while penning peer-reviewed research for scientific journals and tending to day-to-day political responsibilities (albeit not always adroitly). "We [the United States] want to be a leader," Chu told me then. "This is our prosperity."

The Chinese have chosen dams as one way to power their future more cleanly. New dams block the flow of rivers to tame floods and generate electricity. "It is a choice we made without any other choice," says the little old man with white hair who picks at his weathered hands as he speaks with vim. He is Lei Hengshun, a professor of sustainable development at Chongqing University and a municipal government official who believes China will try to rely on hydropower despite the cost in lost villages, even cities, and millions of environmental refugees. The alternative is yet more coal, which is already smearing out the sky, killing people by fouling lungs, and warming the entire world.

There is no easy answer to the challenge of providing power to the poor, but the Chinese have chosen to follow a path laid out by the United States, in particular, the Tennessee Valley Authority: Build large dams to bring power to the impoverished masses and kick-start industry. There is need, there is resource, dams are relatively cheap and perhaps less bad for the environment than all that coal.

That choice by China's leaders has forced more than a million people to abandon their homes to fulfill the dream of a dam under discussion for eighty years that finally became reality: Three Gorges. A new city of Fuling has risen to replace the one drowned as the waters rose behind this megadam, tall apartment blocks rising out of the cornfields and swarmed by construction crews on bamboo scaffolding. Also landslides, earthquakes, foul and stagnant waters, and the end of a river ecology that had survived millennia of Chinese fisherfolk. The baiji finally rendered extinct. Even the forecast has changed. "Because the water is big, the weather change a little bit. More rain and flood," a local government official in Wenzhou tells me over a lunch of noodles. And the local people farther upstream like the Naxi oppose the proliferating smaller dams, which provide little power and serve mostly to enrich local cadres as well as prevent damage downstream at major dams if too much silt builds up.

Three Gorges also provides flood control, the dam built precisely as high (185 meters) as needed to keep Shanghai above sea level 660 kilometers away. Behind the dam sits nearly 40 billion cubic meters of water, a lake that stretches 660 kilometers back to Chongqing, a massive inland city. If the dam breaks, millions downriver will die. As long as it holds, the thirty-two turbines can spin up nearly 100 billion kilowatt-hours of electricity, or the equivalent of one hundred of the largest coal-fired power plants that China or any other country could build. Boats ride low in the water, filled with as much as 10,000 tons of coal traveling upriver, perhaps all the way back to Chongqing now that the reservoir has deepened the old river. Old Russian-made hydrofoils, smoking and wheezing, make for a fast trip up and down the new waters.

The dam seems to employ figures from China's mystical past to prove its mandate from heaven, ancient and probably apocryphal emperors like Da Yu, who similarly tamed this river. As the legend goes, Yu the Great changed the course of Chinese history by placing a mountain in front of the Yangtze River and forcing the water to flow east through the Middle Kingdom, rather than south like the other great rivers that rise in the same Yunnan region. Not to be outdone, the modern Chinese government built Three Gorges.

Guards stand on rigid duty over gravel pits that speak of the massive construction now complete. U.S. companies boycotted the project, but the Europeans had no such scruples, so the local hotels boast television channels in French and German instead of English. Ships slowly rise into view in the locks that raise them to the level of the reservoir beyond. Yichang, the town the dam built, has the feel of a faded boomtown, tattered flags flapping in front of the local branch of the Industrial and Commercial Bank of China, run-down buildings holding closed restaurants, and sealed-up underpasses for traffic that no longer runs.

The Communists even try to control the rain here, flying airplanes to seed the clouds and provoke downpours—a similar effort was made to keep Beijing's skies blue for the 2008 Olympics, with dubious results. "We have taken big floods into consideration," Lei says. "What we did not expect is less water."

At Shigu, a town made famous as the place where the ragged band of Communists led by Mao forded the Jinsha on the Long March that would one day lead them to power in Beijing, the people wait for the next project. So the Hei and Bai Rivers, black-and-white tributaries of the Jinsha, may have their own dams. The goal is to ensure a steady supply of river all the way downstream to the mighty Three Gorges Dam, a completely controlled flow that will destroy the local wildlife, cause Shanghai to slowly sink from a lack of silt, and provide construction jobs and tax revenue to local officials. The Minjiang River alone now has seventeen dams.

All of which is to say that a dam is a complicated thing, much like a city, environmental work, or people themselves. It is in the things people build—roads, dams, perhaps most important, cities—that a better Anthropocene for every living thing on this planet will be crafted, or not. The Three Gorges Dam displaced more than a million people, flooded 13 cities, 140 towns, and 1,350 villages. It has also imperiled the continued survival of the Yangtze sturgeon and the Siberian crane. "I do not just care for the environment, I also care about humans and development," says Cun, who is conflicted about the dams.

The problems of the world can't be solved in a windowless hotel conference room or office. The "we" of humanity is too differentiated. Nor is everyone in those air-conditioned rooms deciding the fate of the world. Nevertheless, that's where the attempt is made, if it's made at all. It is in offices and conference rooms that important people speaking in platitudes have real impacts on the real world, whether that is coltan (a metallic ore used in electronics) wrested from underneath the Congo or a concerted campaign to deny climate change.

There are also un-conferences like Sci Foo that attempt to tackle the big challenges. First held in 2003, the annual event is the kind of place where you can find a Nobel laureate, a tech billionaire, the mad wizard of genetics who looks like a vengeful Old Testament prophet, and a merry prankster all pedaling together on a cycle built for eight that looks like it should go nowhere but actually maneuvers around quite capably. The octocycle was painted in the ubiquitous blue, green, yellow, and red of Google's logo, like everything else on the Google Campus where Sci Foo is held. Participants are encouraged to stow all their devices—to "be present."

You might also catch a Nobel laureate singing karaoke, along with an astronomy wiz not afraid to take her knowledge onto the New York City subway with a sign around her neck, and an ecologist who hopes to log biodiversity into a computer record before it's gone, so that we might learn something from our current mistakes. The physicist who has found the power laws that drive the growth of cities and megacities, Geoffrey West, sits serenely like some desert prophet with windswept hair (even in the climate-controlled confines of the Googleplex) and a bemused smile, as if he were constantly high.

Everyone's name tag has his or her profession, self-declared passion, and only then one's actual moniker. More prosaically, it's the kind of place where you can sit around a small conference room table and muse with professionals about the world in fifty years, the prospects for faster-than-light space travel that rely on "exotic matter" to bend the laws of physics, or the 86 billion neurons in the human brain and when, or even if, that can be simulated in a silicon-based computer.

"We don't need a full understanding to proceed," said that new genetic prophet George Church, and he should know as he proceeds with efforts to resurrect mammoth cells without a full understanding of what's at work. The power of engineering in the face of ignorance is not insignificant; just think of the immune system, which remains incompletely understood and yet vaccines work. "They were awesome," Church said at the Sci Foo I attended. "They're still awesome and we still don't understand."

Simulation may be the wrong route entirely, however. Chemists don't rely on simulation to work with a glass of water and its 10^{22} H_2O molecules. Just shake it up and play with it. Or we can make the equivalent of a human brain, the artificial intelligence that Elon Musk and others fear, especially the "strong" variety that could displace what are known around these parts as "knowledge workers," aka me. On the other hand, strong AI could be the kind of intelligence that makes Einstein's insights look pedestrian and could save us from ourselves, enabling better solutions to the problem of planet-searing asteroid impacts or management of biogeochemical cycles that keep life humming, whether carbon, nitrogen, or phosphorus.

Google seems to be trying to build a world brain of sorts, through events like Sci Foo that create new synapses with the connections between smart people, as well as massive amounts of parallel processing, whether that's just enabling virtual reality to better employ computing and connect people across continental and oceanic distances, or wiring up a world of sensors and drones to keep tabs on the planet. "It's not clear what we should be doing in the world," Google founder Larry Page tells us in his whispery voice. "If you had $50 million, what would you spend it on? Tell me what we should be doing."

Maybe it's printers that can churn out metal or plastic or even biological parts in three dimensions that could enable artisanal manufacturing that allows a country like Kenya to leapfrog the dirty industrialization that has befouled China, but made hundreds of millions of Chinese richer, especially with the clean energy to run those printers coming from bioreactors powered by sunshine. Church and his allies hope to hack the DNA source code of life itself, perhaps enabling a trip to Mars, finally, by story about an ancient ancestor named Ngurunderi. It goes like this: Ngurunderi had many wives and yet he was not satisfied, chasing them continually from morning to night. Fed up, the wives fled, hightailing it across fertile plains to what is now Kangaroo Island, in the middle of broad, shallow Melbourne Bay. But Ngurunderi's wives got there by foot over dry land.

Ngurunderi knew of their treachery and knew where they had fled, so he angrily raised the seas and flooded the plains. He turned these wives into rocks that just barely make it above the tide line between Kangaroo Island and what is now the mainland. Based on the water level, this story is of a time at least 10,000 years ago.

This story is not unique. All around Australia are tales of people who walked to what are now islands, followed by floods, floods that match up nicely to rising sea levels as the last Ice Age ended 10,000 years ago. Nor are these long-lived tales confined to Australia. The Klamath of Oregon have a tale that tells of the last eruption of what's now known as Mount Mazama and its crater lake, and the system of Chinese characters has lasted, transmitting information, for at least five millennia so far. Even the biblical flood may be a millennia-long echo of a time when the Black Sea was a freshwater lake before the rising tide of the Mediterranean spilled over the Bosporus and drowned once fertile lands.

The key to preserving such tales seems to be a cultural focus on truth, not unlike the stories told by science today, accuracy and accumulation. Humanity itself is constructed through such storytelling. Perhaps we are not humanity the wise but rather humanity the storytellers, *Homo relator*, who aspires to earn the title *Homo sapiens* at some point.

We will need better stories for the future: factories that clean the sky instead of befouling it as they work, cities designed to provide homes for people and plants, more educated girls and more women in charge. Imagine a world where solar panels replace chimneys in children's drawings of a house or the bloodthirsty bow hunt replenished buffalo from an electric SUV. The world should be better as a result of our ingenuity.

Entering the Library of Congress is a trip into a repository of accumulated knowledge. Studious faces of scholars past like Seneca stare down from the walls, and the rooms, even the bathrooms, boast marble floors. A Gutenberg Bible, the first printed book, sits enshrined in the foyer. This is a place of reverence for wisdom and a library built for the future.

just sending a DNA synthesizer, 3-D printer, and a power source that c;
build a robot, a cell, maybe even an animal like a human from the atom
up. Call it biological teleportation.

Of course, the people who dream these dreams also dream of gettin
away from the clutter of the old, messy world where the rest of us liv
where they feel strangled by overregulation and burdensome taxes, b
self-report. There are more middle-aged white men than women or an'
other group cutting down on the diversity of ideas and solutions on offer

When incomes and outcomes become unequal, social cohesion breaks
down. I suspect it is this era of declining cooperation, dissension, and
discord from politics to the very methods of educating the next generation that has spawned the present obsession with dystopia. Things seem
to be getting worse because people like me and you have lost the means to
find common ground, or to imagine our way into another's life. Without
cooperation, the complex web of modern industrial civilization will collapse, because it is based on trust. No free market can function without
cooperative structures, and ideological fetish cannot change that.

The Sci Foo-ers are suffused with hope that the bright people of this
world can solve the present array of problems, just as a band of sleep-
blinded villagers can sometimes—just sometimes—fend off a pack of evil
lycanthropes in the game of trust known as Werewolf. At the same time,
Sci Foo can demonstrate the blindness of those trapped in a techno-
optimistic viewpoint, wandering about wearing Google Glasses that
broadcast, even advertise, a Big Brother future with everyone watching
everyone else to no good end.

The primary goal has to be the longevity of human civilization, not
necessarily in its modern incarnation or hunter-gatherer past but some
blend of all the diversity of cultures and peoples, which can provide
us with the options and tools to survive almost any challenge. Not an
Anthropocene, a Democene. That's what brought me to the Library of
Congress, an institution or instrument meant to ensure the longevity of
knowledge via old technologies like this book or new technologies like
an archive of the global conversation on Twitter. Can this knowledge last
as long as an Aboriginal tale about the world before sea level rise? That's
another potential use of this Anthropocene idea, to get people thinking
about longevity.

The Aboriginal band of people known as the Ngarrindjeri have a

Perhaps we should heed the words of Seneca, whose words repose here in his *Consolations to Polybius*: "The seven wonders of the world, and any even greater wonders which the ambition of later ages has constructed, will be seen some day leveled with the ground. So it is: nothing lasts forever." Only adaptable, changeable humanity has that chance.

The future will always be a surprise, just as the unforeseen advent of the automobile solved the horse manure crisis while spawning new problems like the fragmenting of ecosystems by roads or CO_2 from all that fossil fuel burning accumulating in the atmosphere and warming the globe. Every generation will live in special, unique times, like the medieval peasants coping with the Black Death and the creation of the cultural code of chivalry still with those of European descent today. The problem is that the mental architecture that enabled us to master fire for cooking or store food for winter or drought is not as enabled to cope with the long-term planning required to forestall climate change or save species. The resilience of modern civilization is continually tested, and there are as many examples of the collapse of complex societies as there are of adaptable civilizations that persist for centuries or more.

Be very wary of anyone who tells you what human nature is, including me. The only constant is change. We can ignore race or see it. Pursue greed or goodness. Perform selfless acts of heroism or creep in cowardice, often all in the same day. Human nature is a shape-shifting myth. Environments, circumstances and, most important, other people set the boundaries of what's possible within the infinitely malleable human mind. Cultivating kindness may prove key—whether kindness to our fellow humans, animals, plants, or even the land and sea.

Some of us live in an air-conditioned world, lighting set by computers, and with speakers playing the sounds of nature to soothe. Indoor farms and, perhaps one day, whole bubble cities walled off from a world made inhospitable. This is hotel world and the end state of thinking ourselves somehow separate from nature.

Then there are all those folks making day-to-day decisions to try to create a better life for themselves or, more important, their children. That can mean clearing a forest or burning coal. Combine the decisions taken in hotel world with the decisions taken in hovel world and you have the Anthropocene.

There is no simple solution to the problem of humanity. Denizens of all these overlapping worlds will need to strive for an epoch that is less bad than what has come before. Technology can give access to those from hovel world to hotel world, and vice versa. People can change—think of the move away from smoking in the United States or shark fin soup in China. Shifts happen. Human nature is malleable, and we always get the future wrong, like the 1950s engineering fantasies of the world in 2010 that showed modern offices full of gizmos and brilliantined men—but no women.

In the time it took to build both hovel world and hotel world—10,000 years, give or take a millennium or two—the planet transitioned from the frosty depths of an ice age to a long summer that made agriculture more feasible. That long summer will now transition to a hotter, wilder season, and we will have to adapt using the tools at hand. As individuals, humans are almost nothing, even the most wildly skilled among us incapable of accomplishing all the feats completed day after day collectively. The great storehouse of hard-won knowledge—which plant will kill you, how to make a steam turbine, the scientific method itself—is our only bulwark against doom. That many of us use that bulwark for trivial purposes, flame wars via social media, say, only obscures that fact without diminishing its significance. And the bulwark must be tended, whether that is a Wikipedia page or the patient teaching of a child how to farm. Books are a technology to reach across time, just like a culture itself.

If there is a simple solution to our present problems, it might be summed up as: Empower women with clean power.

South Africa is a land of contrasts. Freshly paved roads that dead-end at dirt tracks that lead off into the veldt. Crudely cemented cinder-block homes surrounded by crops across the street from a fenced suburban ranch-style house, complete with car in the drive. Barefoot children carry bundles of branches from the dirt track, across the paved road, and into their homes while the electrical grid ends with a wire running into their neighbor's house. So one family burns wood for cooking and heat while the other enjoys the electrical benefits of coal burned far away.

The most basic reason to want modern energy for all is human health. More than a million children still die from diarrhea every year, a tragedy that could be solved with a little hot water and washing. That's a new benefit offered by the local Aldinville Senior Primary School in the iLembe District of KwaZulu-Natal province. The low-slung brick buildings halfway up a hill provide classrooms for a few hundred local kids, a mass of smiling boys and girls seated at wooden desks in a long line of rooms that haven't been painted in years. The goal is to provide "High quality education to all our learners so that they become accountable, disciplined, creative and marketable citizens of their society," reads the slogan on the principal's office wall.

A metal roof keeps out the rain, but not the heat or cold, which is why most of the kids are wearing sweaters or two shirts on this chilly austral spring day. Water is precious in Zululand, so the rainwater that runs off the roof is collected in large green plastic drums, for drinking and washing. The grounds of the school are dry and dusty, the dirt red and compacted from all those children's feet. And perched atop the sunniest exposure of the rust-red roof is another water tank, elevated on a short scaffold and attached to an array of thin dark tubes. The tubes absorb sunlight and use it to heat the water inside.

The kids call out "soft hands, soft hands," as I walk by outside the classrooms giving high fives. Their smiles are large and bright, and this is as close to being a rock star as I have ever, or probably will ever, feel. My hands are soft compared to their fathers' and mothers' because I don't have to work the fields, just scribble in a notebook and type on a phone. Also because I have near constant access to hot water and soap, and with it the power to wash my hands. That is a use of energy even more fundamental to these students' well-being: hot water. The new solar hot-water heaters perched atop the low-slung tan buildings of this school will offer them a chance to wash up, at least at school.

There's also a hot meal that cuts down on the chance of the kind of microbial contamination that makes people sick. China lacks cold salads in its cuisine because Chinese fields have been fertilized with human waste for millennia. Cooking kills the microbes that might otherwise spread. And China is now the undisputed center of solar hot-water heaters, where every new apartment building sports row upon row and tanks are exported to markets such as South Africa.

China also is a land of contrasts, where a peasant antiquity resides a short drive away from modern, gleaming cities. China represents a possible future for South Africa, or any other part of the world that has yet to benefit from modern energy. If current birth rates hold, Africa will become as densely populated as China and the rest of Asia. But a little economic security is a potent contraceptive, based on the experience of richer countries. Let the poorest nations grow, the people move into cities, and the birth rate will stabilize there too.

As I walk through the classrooms, nervously smiling, overwhelmed by the attention and the hope of these Zulu children, I have only one thought: The world will be a better place because they are in it. A girl in this classroom may well grow up to be the next Ada Lovelace, the original computer programmer, or Rosalind Franklin, who truly discovered DNA's double helix and made George Church's fever dreams possible. There may even be a new Hedy Lamarr here, movie star and inventor of the frequency hopping that makes cell phone communication possible. Good hygiene will help ensure that these girls and boys grow up strong and healthy. More than 6 million children die before they make it to their fifth birthday for want of hand washing. And proper hygiene requires energy to heat and move water.

Each heater tank and its row of tubes can hold 400 liters of hot water, enough for the Aldinville kids to wash hands and eat a hot meal, even on cloudy days like this one. Such a simple act as hand washing could save 600,000 kids from an early death each year.

Then there is the question of light. Light is essential for the students of Aldinville Senior Primary School. Without light, there can be no studying at night. To go from the smell, expense, and danger of kerosene to a clean, well-lit room is perhaps the essence of modernity. And modern energy will help them fulfill the exhortation of the poster taped to the side of a cabinet holding trophies in the principal's office: DON'T QUIT. Don't ever quit.

Back in China, at least some of Cun's efforts have paid off. One family near Liming is trying something new: biogas, at their own expense. Biogas is the fermented remains of human waste and other garbage consumed by microbes in buried pits, "which is good for trees and water," the young lady of the house, an Earth Helper named Feng Yu, explains to me. We chat in her kitchen as the rain falls outside, which doesn't stop the chickens from clucking and chirping and stalking about the courtyard. The club has recently been teaching her about the conflict between the dwindling number of Asian black bears in the forest and the local humans and, in particular, how to avoid that conflict. It is hard to draw her out on any subject. She is clearly nervous talking to a foreign journalist, skittish like a foal, even if her dream is to perform one day, maybe as a dancer. She flips through her blue honors book for me, which lists her many achievements: a first in essay, third in painting. "Everybody loves it," she says of the club. "It's fun and they can contribute to the hometown."

She herself loves the remnant forest of tall, old trees. "When you're tired, it restores you," she offers, refusing to use her recently learned English in favor of Mandarin. The green leaves on the trees are a symbol of hope to her. Plus, the forest is eminently useful for keeping the soil around and controlling floods. "If we overuse the forest, we will have a bad impact," she adds after thinking for a moment. "Use as little as possible."

"But how do you convince people?" I ask, as we watch the chickens knock corncobs around in the courtyard.

"Use [your] knowledge to pass to more people," she says, after a few more moments of reflection. She is the first girl in her family to progress so far in school. "To protect the environment is to protect human beings."

Take, for example, that biogas. Not only does using methane produced from their own waste save wood, and therefore the forests, it also saves her parents. "Parents do not work as hard as before," she offers, "and it is also a benefit for the forest." The kitchen still smells of decades of smoke, but now there are two burners connected to a thin tube that carries the biogas in from the pit outside. In this shady, cool valley, the microbes cannot make much biogas, but it is enough for two meals, Feng says, cooked over a smokeless blue flame.

The best way to spread the word, she thinks, is art. Maybe she will make a dance performance herself one day, to give a voice to the animals when they get hurt by humans. Animals and humans are equal, she thinks, and the animals should have room in the forest too.

Feng was born into a China different from the one her parents knew as children. There are more opportunities and more freedom, and a big reason for that is education, particularly of young girls who then become empowered women like Cun Yanfang. The Age of Man could continue to exclude half the "we" that makes up humanity, namely, women. This age could be too well named since patriarchal societies have been the cause of most of the big changes that mark its beginning. The Anthropocene moniker also nicely plays up our anthropocentrism, excluding all our fellow travelers, like the golden monkey and yew tree.

The world gets better when women are empowered rather than marginalized: More economic growth, better health, less environmental destruction all go hand in hand with free women. The statistics show this again and again; the French philosopher Charles Fourier was right when he noted in 1808 that the best countries have always been those that allow women the most freedom. The best and simplest way to change the world for the better is to educate girls. It is a simple fact, widely ignored, that must be embraced if an enduring civilization is the goal—and one that many have recognized over the years, including the great inventor Nikola Tesla. "It is not in the shallow physical imitation of men that women will assert first their equality and later their superiority, but in the awakening of the intellect of women," he told *Collier's* magazine in 1926. "Women will ignore precedent and startle civilization with their progress."

The peasants of Yunnan farm in a way that has not changed much for hundreds of years, including backbreaking labor to till, plant, harvest, and transport crops from small family plots. Old women and men still carry huge baskets of crops or firewood on their backs, and from the air the tiny fields look like some highly refined, intricate precursor to the computer circuit board. It is such agriculture that may have given us the sociality—like bees or ants—that then allowed this naked ape to conquer the world.

"Local people, their voice is never heard," says Li Shiyang of the international environmental group Rare, who teaches teachers how to train fisherfolk to fish with nets rather than dynamite and employ biogas rather than wood for cooking fires. "Usually conservation experts go there and tell them what to do." And that doesn't work, because it fails to take into account local knowledge as well as local needs. And these local people must be invested in the project, paying something toward that new biogas system, or they will never take it seriously.

So, the experts like Cun and Li participate and teach but do not dominate, allowing minds—and practices—to change, which is where Cun's intimate knowledge of the local community is indispensable. Without her, nothing is possible. "What we're good at is influencing local people," Li says. "For the local villagers, for them the most important thing is not the monkey, it's how they can make a living. If you preserve the forest, there are more herbs and mushrooms," which means income for the villagers and, indirectly perhaps, preservation of the forest and a home for the monkeys. True solutions benefit both people and the planet.

Of course the forest still suffers: "We farm anywhere we can farm," Cun notes. Yunnan is beautiful but poor, and thus must find a way to eliminate poverty and conserve nature. The empowerment of women is exactly how those twin goals can be met.

Meanwhile, the mountains have been covered in a new pine monoforest while, down in the valley, fast-growing eucalyptus trees line the road. As the Chinese proverb goes: "Planning for a year? Sow rice. Planning for a decade? Plant trees. Planning for a lifetime? Educate people." Especially those people most often left out, like Feng, a daughter of peasants.

The children of Yunnan now learn on donated computers that give them access to a wider world, despite the Great Firewall of China and in accordance with government mandate. Feng and her friends may be shy and even poor, but education opens opportunities for them that were closed to their mothers. Weibo, WeChat, and QQ spread information, context, and contact.

When discussing a controversial subject, like whether or not a Chinese government—municipal, provincial, or national—will live up to its promises, it is better to circle in slowly than to question directly. To illustrate, Cun drew me a circle on her palm that spiraled inward compared to a straight line. "Environmental protection is our national policy . . ." she noted, before trailing off. In other words, that doesn't mean the environment is protected, just as women don't necessarily get their due.

Part III

A Better Anthropocene

Chapter 6

City Folks

As for those who would take the whole world
to tinker as they see fit,
I observe that they never succeed:
For the world is a sacred vessel
Not made to be altered by man.
The tinker will spoil it;
Usurpers will lose it.

—LAO TZU, *TAO TE CHING* (CHAPTER 29),
TRANSLATED BY RAYMOND B. BLAKNEY

Fan Changwei keeps his silence mostly about his youth, but he will offer that he spent three years in the countryside. The middle-aged lawyer, who has worked his way up the leadership ladder at the local Environmental Protection Bureau, was sent down to Shijialing village, like many of his urban compatriots before him during the Cultural Revolution, including the man who is now president of China, Xi Jinping. As soon as possible, however, Fan came back to Rizhao, the city where he was born and raised. "I like the small city. It is not as crowded or polluted."

The city of Rizhao faces South Korea across the Yellow Sea and, beyond that peninsula, Japan. The resort town takes its name from an old poem—*ri chu xian zhao* (first to get sunshine). These days, 3 million citizens enjoy that early sunshine, along with the beach, sea breezes, and a swelling array of amenities like modern apartments outfitted with televisions and hot water. The men and women who govern Rizhao have another hope for the small city too: to make this city one of the first in

China and the world to spew no more greenhouse gases through coal burning or the internal combustion that drives cars than the city can get rid of through technology fixes or planting trees.

"The city will try to go carbon neutral," says Fan, who has helped lead this seemingly quixotic effort. He communicates with me using his own imperfect English that makes it difficult for him, he says, to express all he is thinking. All I can think is that it is far, far better than my own few words of Mandarin. He resembles one of the tall, thin smokestacks on the local coal-fired power plant, only he burns through as many cigarettes, instead of coal, as time permits. Taking a deep drag and exhaling, he adds, "I don't know when we will succeed, but we will move in that way."

Rizhao is just one of a handful of places globally—Copenhagen in Denmark, Arendal in Norway, Melbourne in Australia, and the entire country of Costa Rica, to name a few—that even attempt to balance their daily exhaust of carbon dioxide, the same greenhouse gas breathed out by Fan amid all that cigarette smoke. The local Communist cadres who run this city believe that leading a transition to a cleaner economy will help drive economic growth, and their careers. Rizhao is growing, like most cities in China, stamping out skyscraper after skyscraper and apartment block after apartment block amid the long, low buildings that make up its older homes and businesses. The local mayor went on to become vice governor of the entire province of Shandong due to his success in creating a circular economy, which is Rizhao, and China's answer to the challenge of climate change and perhaps a way to build an enduring Anthropocene.

"To save money is very important," Fan notes, as an explanation of sorts.

The circular economy is a simple idea, hard in practice. No waste. The residue of one process, say, making citric acid for soda at the Jinma Industrial Park, turned into the feedstock of another. The grainy smell of sour mash wafts throughout the park, and a whirring roar fills the air. The key component is a special "biodigester" from Holland, essentially an industrial stomach for turning corn and sweet potatoes into citric acid, stinky water, and residual grains. The wastewater pools exude methane, most to be captured as what the Chinese engineers call marsh gas. This industrial stomach makes 25,000 cubic meters of marsh gas per day, and there are ten similar enterprises throughout the small city. The marsh gas gets burned to run six generators pumping out electricity. The residual heat from that process is used to warm local buildings. As for the leftover bits of corn, sweet potato, and cassava, that becomes feed for the animals.

In the village the old authorities sent Fan to decades ago, those animals include minks kept in tiny cages for the fur trade by farmer and entrepreneur Shi Tongkang. The minks' waste, and the waste of Shi's family, goes into a pit not unlike Feng Yu's, far to the west in Yunnan, a smaller-scale version of the industrial stomach. In the pit, microbes break down the waste into water and methane, and the gas is captured, fed through a thin tube, and routed to the small burners of the Green Lilly stove in the tiny tiled kitchen. The resulting blue flame makes Shi's kitchen one of the cleanest in the Shandong countryside, Fan suspects, at least when it comes to the indoor air quality that kills millions around the world, more than AIDS, malaria, and tuberculosis combined. Similar systems help households fueled by pig shit or fox dung, more than eighty in all in the villages surrounding Rizhao, encouraged by government subsidies that cover at least half the cost of installation of such biodigesters. And some of the marsh gas is burned in the ubiquitous greenhouses that grow vegetables, both to keep the temperature warm and to kill off pests. In the future, the hope is to make so much of this marsh gas that it can even be used as a fuel for the vehicles multiplying on Rizhao's streets. "The technology of compressed marsh gas is not too difficult," Fan says, but it is very expensive. A more efficient stove alone will set a peasant back at least 1,500 yuan (about $230), a significant portion of a peasant family's annual income.

The circular economy is an old idea in China, as old as the human dung gatherers who sold night soil from the cities to the farmers as fertilizer for the fields. "Organic fertilizer is much better than chemical fertilizer," Fan notes, adding that it is used to grow the region's "famous" tea, some of the northernmost tea grown in China and another example of how human desires reshape entire regions of the planet.

If returning to this village bothers Fan, he doesn't show it, and he just shrugs at my questions about it and takes another drag on his cigarette. In the 1960s, Mao deported millions like Fan from cities across China, and the urban population was cut nearly in half for a time. By the 1980s, those government controls eased and a flood of peasants, by some estimates as many as 100 million people, provided the cheap labor of the Chinese economic miracle. That labor is kept cheap by denying these entrepreneurial peasants any rights under the *hukou* household registration system that keeps people literally in their place.

Fan seems truly unperturbed to be back, guiding me from his old school in Xihe to another farmer's courtyard, standing around in his blue suit, white shirt, and glasses, smoking, as peasant women in brightly colored smocks gather in the street to cluck at the foreigner in their midst and then explain a little bit of the workings of their new, more environmentally friendly lives. Wearing a suit in a village is ridiculous, so I cannot be sure if the women are smiling at me or at how foolish I look. Lush wheat fields surround this village in spring, and the ginkgo trees, planted for medicinal purposes, Fan tells me, have sent out their fan-shaped leaves. The favorite tree of the modern Chinese Communist cadre—the poplar—easily outnumbers the useful ginkgo. The poplar is popular because it grows straight, tall, and fast. Unfortunately, some of those poplars planted with military precision have been tilted by storm winds and now list like drunken soldiers. Buried in the back of this village's poplar forest are tall mounds with pots on top: tombs of the ancestors. It is not hard to imagine that every inch of this ground has been a burial site at some point in the thousands of years it has served the Chinese people. The only wildlife to be seen are the mosquitoes that hide in the shower drains, the few magpies in the heavily dusted trees, and the foxes and other animals kept in cages by peasants running a fur-farming concern. The elephants that once roamed much of China have been gone for centuries, their only legacy a fondness for ivory that is killing off the remaining elephants everywhere else in the world.

The houses here in the village look like shops, with aluminum-paneled windows that reveal sitting rooms and bedrooms. Roofs are thatched, and the best houses are really brick-walled compounds. Hand-woven mats are unfurled each night and placed over the greenhouses to keep the heat in, then rolled up again in the morning. Life in the countryside is labor-intensive. Pesticide spraying, raking, burning crop residue—all these tasks fill farmers' days even though they don't own the land they tend. The burning adds to China's haze, but it seems no government can stop it.

The circular economy is merely a synonym for thrift, an old-fashioned value everywhere in the world. After all, the poor don't produce much trash, as anyone who lived through the Great Depression can attest. "Economics are important," as Fan puts it. In this sense, aluminum cans may be the ultimate object of the circular economy: endlessly recyclable with no diminution in quality and saving much of the mega-energy required to create aluminum from bauxite ore in the first place. Everything else is perhaps a bit more complicated.

So, people like Fan drive Rizhao toward carbon neutral through a mix of using less energy to produce more, recycling the wasted energy of one process to drive another, and getting energy from cleaner resources in the first place—replacing coal with solar, wind, marsh gas, even nuclear. The small city's economy is cargo port based, but also manufacturing: electronics and electrical systems for cars, furniture making, papermaking with its soured perfume, textiles, and food processing.

Fan himself is living the Chinese dream, which is perhaps why his face crinkles in a seemingly perpetual smile when he talks. He once walked or bicycled to work to be personally carbon neutral but now has access to the bureau's fleet of cars and drivers—and his own Volkswagen at home, which he usually lets his wife, Fang Wang, drive. The couple's life revolves around their daughter, Fengyuan, who has dreams unimagined by her father during his own schooling in the village, including perhaps studying abroad one day. The Fan family recently moved to a new home near the seashore, not for the breezes, though those are nice, but to be close to her high school. Still, Fan did make sure his new home had double-paned windows and plenty of foam insulation to improve its energy efficiency.

His workday consists of enforcing the law, such as fines for companies that pollute too much, maybe even shutting them down. But it is all within reason because helping peasants deliver themselves from poverty via the city remains China's overall strategy: political stability through economic growth, or the government getting out of the Chinese people's (economic) way. On this day, though, he picks me up from the airport in Qingdao early in the morning after my flight from Beijing. It is not sunny in the port city as I fly in; perhaps it is a mist from the sea or rain or yet more haze, occasioned by a construction boom that produces buildings stamped out like LEGO blocks, and impenetrable.

I get into the back of the bureau car with Fan, and we have a halting chat for the more than an hour's drive to Rizhao, passing shrimp farms and industrial plants. The driver, who refuses to give his name, served in the Chinese army for five years, before landing this gig. He races the car as fast as he can, touching 160 kilometers per hour at points. The newly planted poplars lining the highway expose the yellow earth of this part of Shandong Province. "Trees are also good for carbon neutral," Fan observes, smiling, though government officials had to learn the hard way that planting all the same trees has left them vulnerable to death by single pest. Now the government plants a variety, albeit still in the same rigid lines easily distinguishable from the random scattering of a more natural forest.

People have eaten away at the hillsides here in the quest for ever more material for concrete, earth movers and mechanical shovels digging into the mounds to make smooth roads or concrete foundations elsewhere. Geologic features seem to be collapsing in on themselves, and there are no trees on those distant hills. Erosion is everywhere. The local crops are mostly wheat and peanuts, Fan says. Everything is cultivated in tiny, human-size rectangular plots. It is the amount of land a single peasant family can work and the hallmark of this deeply worked-over landscape. Lonely smokestacks poke up out of the fields where peasant kilns churn out bricks. In all the kilometers between Qingdao airport and Rizhao, there is not one centimeter of land unmarked by human hands except perhaps some of the stone outcroppings on the mountains.

As we travel along the new superhighway, an overloaded truck trundles past us, jingling a bit as it bumps down the road, filled with green glass collected by people paid to pick through garbage and collect bottles. The trucks, no matter what they carry, are open on top, at best covered with a tarp or netting. "They need to make more money," Fan explains, so they take the risk of spilling an oversized load or, worse, tipping the truck. We overtake a tractor creeping along the new road as fast as it can.

China has become an enthusiastic participant in the current globalization, kicked off by the British Empire and brought to its current state of interconnectedness by the United States. The road signs on these gleaming new highways are in both Chinese characters and the Roman alphabet that spells out English words. Today's Chinese children are as fluent in English as their parents were in Russian in a more ideological but perhaps no less divided world. This same globalization makes China's air pollution—whether the carbon dioxide that invisibly warms the globe or the tiny particles of soot that smear the sky—some of the worst in the world. On certain days, the air in Beijing has approached 1,000 points on the air quality index, a measure of air pollution concentrations over time and, essentially, an off-the-chart level. In the United States, even a forest fire combined with smog from cars and haze from power plants rarely sends levels as high as 300. This is the world-famous "airpocalypse."

Athwart that pollution apocalypse stands Fan. For years before carbon neutrality became the hip new thing in Rizhao, the goal was to tamp down on dust and sulfur dioxide, and eliminate water pollution. Fan and his colleagues have so far failed to do much of that. "It is not easy," Fan says of environmental protection in general in a land trying to grow as fast as possible into an economic giant and a global superpower. "We will try. And manage to do . . . something."

The city is humanity's greatest invention, the hive mind from which other inventions spring. The city grows like life itself, following the scaling laws that govern the growth of galaxies. Cities are adaptable and changeable while also offering a certain stability, a fount of random encounters but also capable of being mapped and searched. And cities can have a remarkably long life span. Hiroshima and Nagasaki, for instance, came back even after being hit by atomic bombs, and continue to thrive.

The city may even have preceded the settled agriculture seemingly necessary to make a city possible, judging by what the archaeologists have found under the mounds at Çatalhöyük and Göbekli Tepe in Turkey. The city may have spawned the agriculture that conquered the world through inhabitants keeping wild grains to eat over time and replanting some of them, slowly developing domesticated grains like wheat.

Contrary to modern mythology, it is not clear if the city first sprang from a gathering of farmers, trading the fruits of various labors. In that myth, the city has economics at its heart, just like the megalopolises of the twenty-first century. But while it is certain that trade is one important reason for cities to form, the idea of the city may have first sprung up around a place of devotion, some arcane religion bringing people together in common worship and cause. Modern cities still build such a shared identity—just ask the fans of any local sports team. Or it may be that cities formed at places where bands of hunter-gatherers met to feast and swap tales of prowess. The city may then spring from ritual or ceremony and the drive for status. Or it may be all of the above, or none. As the British archaeologist Ian Hodder has written of Çatalhöyük, "One of the main puzzles is why people concentrated together at all."

Regardless, humans do it again and again. The city almost seems to be an outgrowth of our biological heritage, though it took at least 180,000 years for the first known cities—towns, really—to emerge after *Homo sapiens* appeared on the scene 200,000 years ago.

The future will judge us by our middens, just as modern archaeologists attempt to glean clues of the past from the middens of our distant ancestors. The modern city of Rizhao is not too far from one of the oldest confirmed settlements in Asia: Liangcheng, a Neolithic village 20 kilometers up the coast. The settlement is approximately 4,600 years old. Buried beneath the topsoil there lay the remnants of long use, including the highly polished black pottery shards of the so-called Longshan culture, which flourished in the Yellow River valley and built great cities with earthen walls surrounded by moats. This culture collapsed roughly 4,000 years ago for reasons lost in time. What remains clear is that China required centuries to recover.

At Liangcheng, perhaps the people drank *huang jiu*, a yellowy brandy made from rice and reputed to be the oldest alcoholic drink in China, found in Neolithic tombs as residue at the bottom of pottery jars. Still made today, the yellow wine is considered to be medicinal, good for getting the blood flowing, like a massage or an endless series of shots of more common rice wine to the shout of *Gam bei*, Drain the cup, with a Chinese official.

Certainly, where there was farming, and hence excess energy provided by abundant crops, cities grew. Wheat seems to have enabled people to build the world's first true cities in the Middle East—Ur, Uruk, and Eridu in Mesopotamia—at least as far as modern archaeology can tell, though undoubtedly there are more secrets that will upend this view to be discovered under our feet. At any rate, around the same time, early cities arose in the valleys of the Indus, Nile, and Yellow Rivers. A little later, people began to build cities in the Andes of South America as well as the lusher regions of Central and North America. The idea of the city grows easily from the human mind, just like the beehive or anthill for the other eusocial species on the planet.

For roughly 10,000 years, in the region where the city of Damascus stands today racked by conflict, there has been a city of some form or another. Similarly, Athens has been inhabited for 7,000 years. Yeha in Ethiopia has been a city for at least 3,000 years, and Cholula in Mexico has existed—waxing and waning—for more than 2,000 years. After fire and farming, it is cities that remain our most enduring creations, suggesting that the Anthropocene might be better named the Urbocene.

The twenty-first century of the Christian calendar heralds a new era of city building. For the first time in recorded history, more people live in cities than in the countryside. In a small circle that encompasses China, India, Japan, and Indonesia, more than half of the world's 7 billion or so people currently reside, and they live more and more in Delhi, Jakarta, Singapore, and Tokyo, as well as a host of smaller cities whose names remain unknown to the wider world, like Surat and Bandung. This second outbreak of the city is nowhere more obvious than in China.

The Middle Kingdom has boasted major cities for at least 5,000 years, but even as recently as 1940, China had fewer than 70 cities in total, and no skyscrapers. In 1980, the nation's cities contained 200 million people. Now the rapidly urbanizing country has more than 700 cities of varying sizes that are home to more than 600 million people. The national bird has become the construction crane and new cities mushroom from the coastal plains of Shandong to the mountains of Yunnan as migrants leave farms and pursue new jobs in the burgeoning factories.

China's modern story is part of a new way of life. This novel story started roughly 200 years ago in Britain, spread to the rest of Europe and then the United States, and now finds itself being repeated in the titans of the rapidly developing world—Brazil, China, India, Turkey. It is the story of industrialization. Such city-born industrialization has helped take people's impacts to a planetary scale. People now move more earth to build cities and infrastructure than do all the world's rivers and streams, for example. We've made more than 500 billion metric tons of concrete, much of it for our cities, and enough to put a kilogram on every square meter of Earth. At the same time, cities pack people together, which means both more ideas and fewer environmental impacts. Such intense use of a limited space to house and feed billions may prove the archaeological signal of an enduring human epoch, though the Anthropocene may prove hard to find in the far future. Urban land covers only 3 percent of Earth's landmass.

To house everyone, however, people will need to build as much city—

buildings, sewers, transportation, power—by 2050 as people have managed to cobble together in the last 4,000 years. How that city building gets done could ensure a world racked by pollution and weird weather—or not. China has embarked on a building spree, with local governments enriched by pell-mell development that has seen more than 2 billion square meters of apartments and offices constructed in just the last few years.

A single character dominates Chinese cities: *chai* (柴), meaning "demolish." I used to see it scrawled on the walls of ancient *hutong*s as well as 1990s-era high-rises in Beijing, though these days it is more rare as the demolishing continues. More than a million of the poorer among Beijing's and Shanghai's residents have been relocated from their homes to apartment complexes to make room for more lucrative skyscrapers, luxury malls, and new roads. The pace of building these luxury apartments far outstrips the pace of "affordable" housing. In this corner of hotel world, the rich live north of Beijing in gated communities with names like Orange County and Long Beach, while the poor concentrate in the south and see their hovels bulldozed in the name of new development. China has become "*chai-na*," in the words of the cultural critic Sheldon Lu, the place of a frenzied "tearing down," partially to build and rebuild up.

In China today, single-family homes connect by arterial roads to major highways whose entrances are guarded by the gas stations needed to fuel internal combustion engines, all of which sprawls across former farmland—much like the United States in the latter half of the twentieth century.

This is not the kind of intensification that will ensure an enduring civilization, in China or anywhere else. Smarter cities may help, wired to record their own metabolism and fix it quickly, whether traffic congestion or a blackout—and will require smart people to run them. Above all, the power to run cities smartly and efficiently must come from clean energy. But the city is perhaps too organic to ever be truly smart in the ways dreamed of by engineers.

From space, the profusion of cities lights up the night. That picture is also one of waste—light pointed up rather than onto the roads or buildings that require illumination. As a result, mountains of coal get burned and smokestacks spew pollution into the sky so that the stars can twinkle down unseen above a record written in light of a wasteful civilization. At the same time, all that light enables people to learn the past and invent the future.

Cities now encompass multitudes, whether the gridlike rigor of a New York or the concentrated poverty of a Mumbai, where an army of workers dwell on a mere 10 percent of the city's space and issue forth to beat the laundry from high-end hotels and homes and ensure that it is delivered the very next day, crisp and fresh, to rich guests and residents who take up the other 90 percent of the land. New Delhi's air is at least three times worse than Beijing's. Both rich and poor breathe that same air, but walled compounds and walled neighborhoods protect a middle class from their impoverished brethren. The Indian middle class's main function, similar to the middle class's role the world over, seems to be to consume. The bazaars in the subcontinent, with their dark alleyways and dirty shops, clutter everywhere along with innumerable little fires, and have become a bootlegger's dream, full of pirated computer equipment, tools to flout governmental dreams of control. Malls are just bazaars writ large.

The future is built on the past, just as a mosque in Lahore sits next to a Sikh shrine. When the Sikhs controlled this city, now in Pakistan, they used that fort as a stable, and iron rings remain in the outer walls. This still rankles. Now the Muslims repay the favor, allowing the Sikh shrine, once covered in teardrop and oval mirrors that reflected and multiplied the faithful as they moved through, to decay and collapse. Then again, so does the fort itself, as bustle, commotion, and poverty rise up around it and kites float like ash in the sky above. Cutting one kite with the glass-embedded string of another so it flies off or drops like a stone, but in either case is lost, teaches children the ethos of the city: Take care of number one and damn the others. Food, space, money, water, any resource—competition is everywhere in the modern city.

The Sikh shrine and Mughal mosque sit within a fort built by the last of the great Mughal emperors, Aurangazeb, who ruled an empire that included many parts of modern-day Pakistan and India. He built a whole city, named (what else?) Aurangabad, in the state of Maharashtra, near the southern limits of his power on the Deccan Plateau. A mere town of 75,000 inhabitants when India earned its independence from Britain, the city now exceeds a million souls.

Everywhere cities grow—from Mexico City to Dhaka—and too many ideologies compete for space. Religions lurk behind every corner, like the colorfully bejangled Tata trucks named St. Peter and St. Paul that narrowly avoid collision with Ashok Leyland trucks named Shree Ganesha and Allumkadevi in the south Indian state of Kerala, Land of the Coconuts.

Modern Beijing is burning, its sky choked with smoke, turning an afternoon into dusk, the tang of combustion even filters into the airport terminal shaped like a mythical dragon, no doubt a fire-breather. The sweet smell of diesel and gasoline in the parking garage comes as a relief in a city that has existed in one form or another for at least 3,000 years.

The tang from all that pollution may be the taste of development and the peach haze of a morning the look of progress, the price of an economic project that has enabled hundreds of millions of peasants to lift themselves out of lifetimes of poverty. In 1981, the majority of people in China lived on less than $1.25 a day. By 2008, some 660 million Chinese—or two times as many people as live in the entire United States—had worked their way out of such extreme poverty, helped by the blind eye of a nominally Communist government and their own pluck.

My flight from the United States to China has been a bit of time travel and much more than the crossing of the international date line. I have traveled more than a half century into the past, when a killer smog killed twenty people in Donora, Pennsylvania, and the skies of Pittsburgh were dark at noon. If there is another way to grow rich, China has not found it. China is currently living in the future past, a Dickensian steam punk sci-fi drama in Mandarin, complete with high heels and disfigured orphans. Life goes on in Beijing under this empire like the others before it, taxi drivers gambling and laughing over cards, old people murmuring as they shuffle mah-jongg tiles. Kids play in the streets, a gang of small boys wrestling and already smoking cigarettes in the *hutong*s. The difference is the power lines leading into the *hutong*s and the brilliantly painted majestic doors, even if weeds poke up here and there among the roof tiles of the rapidly gentrifying area.

The business of any city is business. On the industrial outskirts of Beijing, a sprawling factory of 15,000 square meters churns out blades and other components for monuments to the need for clean energy that tower over the blustriest countryside. Sirens sound as cranes prepare to lift wind turbine parts from one place to the next. The Goldwind (Jin-feng) factory can build 800 giant wind turbines in a year and has been churning them out since 2006 based on "digested" German technology (that is, copied but with subtle differences). The company got its start far to the west, in Xinjiang, with imported Danish turbines, and since then favorable winds have puffed up the company into a global colossus with turbines spinning on every continent except Antarctica.

Fiberglass and cast steel come together in the laser blast sound of a bolting machine while workers in face masks shower sparks as they weld pieces into place. Copper coils the size of giant witches' cauldrons wait to be employed. The air stinks of chemicals, oil, and ozone as generators are tested, but there are also whiffs of the sweet smell of welding. It is coal burning that makes this factory for wind turbines hum, the riddle of renewable energy self-replication unsolved. And it is corruption that created this golden wind in the first place, connections between private enterprise and the provincial government in Xinjiang, a warren of hidden subsidies, including alliances with state-owned power companies and central government requirements that 70 percent of such equipment must be produced in China. Energy mercantilism may be an economist's sin, but it is the reality of this world.

Business and capitalism took Americans from living in log cabins and dying at forty to living in sprawling ranch homes with a car (or two) and an abundance of other material riches. In other words, industrialization took the United States from poverty to wealth. At the end of that road, 80 billionaires, a group just twice the size of my graduating high school class, control as much of the world's wealth as the bottom 3.5 billion people do. They are also the people most responsible for the damage all around us.

It's not a matter of whether the Chinese will adopt the American life-style with its cars and suburbs—they already have. I meet a campaigner for the environmental group Greenpeace in China who comes from a relatively well-to-do family. "I am quite ashamed because my family has two SUVs," she tells me, refusing to give anything but her English name, Sarah. She explains, giggling nervously, that her sister just bought one to provide a sturdy and safe car for her new baby.

I visit the Greenpeace offices in Beijing, a warren of cubicles filled with chattering young men and women. The hallways are dim, lit by the wan glow of compact fluorescent lightbulbs, and the elevator lacks buttons for floor 4 (a number associated with death in China), 13 (bad luck for European types), and 14 (too similar to 4). What a city like Beijing saves through dimly lit corridors is more than squandered on brightly lit skyscrapers and neon tableaux, just as in the city that never sleeps: New York. As Sarah notes, "There are so many things we have to solve at the same time." Air and water pollution, poisoning the land, feeding the people, ensuring enough clean water, combating climate change.

Volkswagen Santanas clot the roads in megacities like Beijing and Shanghai. These cars have driven China from self-sufficiency in oil to its current status as the world's fastest-growing importer, and therefore coddler of dictators whether they be in Sudan or Saudi Arabia. But all that oil use means the Shanghai lifestyle is broadly similar to that in New York or London. As some foreigners put it: Shanghai is not China, Shanghai is easy.

Shanghai is a younger city than most others in China, growing to prominence only in the nineteenth century, but it has been richer longer. The biggest city in the most populous country in the world, Shanghai is global capitalism with Chinese characteristics, an amalgam of Western ways through its colonial history and China's great modernization project, kicked off not far from here in Anhui Province by a smattering of farmers who returned to farming individual plots. Today in Shanghai, traditional winding hutments fall before massive new apartment towers, but the sycamores, called "French trees" by the locals, are preserved to line the streets.

China's is a vast urbanization, affecting cities both large and small as young people flock to the metropolis for work or new lives. There are as many as 200 million new city residents now, and some of the new megacities that result have yet to reach global recognition. Chongqing is one of the seven furnaces of China and plagued by pollution. There is no horizon in this gigalopolis of 30 million people, the view obscured by the haze. The sky is a grayish white in all directions, the baleful influence of the sun on all that smog and soot.

Chongqing has stood for more than 2,000 years above the confluence of the Jialing and Yangtze that forge a river worthy of the Chinese name Changjiang (Long River), which flows all the way to Shanghai. Here it is boom times for capitalism. China Western Power Industrial Co., Ltd., offers power stations, black liquor (waste from papermaking) recovery, biomass, refuse, blast furnaces, gas boilers, and valves via ads placed on the headrests of seats on Sichuan Airlines' flights to the hinterland's commercial capital. Chongqing hosts factories for cars, motorcycles, power stations, papers, and chemicals like pesticides and steel, like an entire Rust Belt crammed into one city. It even looks like Pittsburgh before the steel mills came down to make parks, with its two rivers, hills, and at least twenty-eight bridges. The people drink from the Jialing, the same river where pesticide and papermaking factories dump their waste. Water treatment equipment is installed but turned on only when government officials check (or foreign journalists visit), a problem common also to power plants and their air-pollution-control technology.

At the same time, thousands in Chongqing and across China still live on less than $100 a year, a reserve army for global capital. These workers roll up their shirts to expose their stomachs in an attempt to beat the heat, along with rolled-up pants or shorts. The pollution used to be even worse, according to locals, the sky green, no sign of the stars at night, and a killer miasma over the entire city. Now there are at least occasional "blue sky days," as the central government calls days with less air pollution. But the air just outside Chongqing is immeasurably worse since the local government relocated the most polluting chemical factories to the outskirts of town, amid farm fields.

Chongqing, like Beijing, has its suburbs, gated communities of condominiums with guards, streets lined with the same French trees as in Shanghai, and domestic workers squatting outside the gates. It looks like California down to the palm trees and adobe villas. There is at least one car in every driveway, and car dealerships line the roads just off the highway: Lexus, Volkswagen, Kia, Honda, Hyundai, Suzuki. Buy local is not part of Chinese domestic politics yet. Even the taxis are Hyundai Elantras.

There is a real Rust Belt in China, the northeast and cities such as Shenyang, originally built up industrially by the Japanese as Mukden, part of their puppet kingdom Manchukuo. This city is more than 2,000 years old and the air here is crisp and cold but rancid, like the smell of sewer gas that leaks out of shower drains in otherwise palatial accommodations at hotels where the conference rooms are named after U.S. states. Once a foundry of Mao's revolution, this city of more than 4 million residents now struggles to reinvent itself as the old monolithic state-owned enterprises are shut down or transformed. Maybe that reinvention can come from the wind turbines made at the Lucky Wind "new energy site" in an industrial park rising from the mud on the outskirts of the city, capable of churning out 1,000 A-Power windmills a year at $1 million a pop for assembly and installation, though that is covered by a twenty-year lifetime guarantee (presuming the company survives that long). Most of the company's products have been installed in China, though some have been exported to Germany, thanks to the low, low prices and government ties. The flags of China, Germany, and the United States flutter over the gates, and the walls of the boardroom are covered in pictures of the founder with former House Speaker Nancy Pelosi and former leader of the Senate Democrats, Harry Reid. On the factory floor, turbines for the booming wind market in Texas are being assembled, and even the slogans are in English: DETAILS MATTER. Blades the length of ten cars await shipment in the parking lot outside the hangarlike factory. These wind turbines go on to compete with Goldwind's turbines from outside Beijing and the hundreds of similar manufacturers across China. Chinese companies copy not only foreign companies, they also copy one another, whether it's a glut of solar panel producers or steelmakers.

Wind turbines don't find a comfortable home—yet—amid the Chinese grid in the northeast, which is run out of a modern palace with a marbled entrance hall guarded by actual soldiers. The old industrial heartland of the northeast now has more power than it needs, so it ships the electricity south to more populous, booming provinces like Jiangsu near Shanghai, or just fails to hook wind turbines up to the grid at all, spinning in the wind to no end. Most of the power still comes from burning coal, like the new Dandong Power Plant near the coast, running at diminished capacity due to the global Great Recession when I visit. The plant itself saves coal by running almost no lights inside except in the control room where workers monitor screens all night, including a televised image of the pulverized coal being fed to the flame.

Maybe the knockoff Toyota minibuses churned out by Brilliance automotive at a full-fledged worker-factory compound, including dormitories, will provide local economic salvation. It's the full company town model. The old state-owned enterprise has converted to private hands via the local government and hopes to make a killing off of digested Toyota technology. Even the Brilliance trademark is merely a reoriented Toyota one. Fortunately for customers, perhaps, the engines are still made by Toyota. The factory reeks of machine oil and rubber. The whirring of drills fills the air as car chassis drift by slowly on an overhead conveyor belt. The slowest worker sets the pace, and the management here seems nearly as grim and sleepy as any old-time Detroit auto boss. Faster workers stand idle, checking cell phones as gaps develop in the line that churns out 320 minibuses per working day, nearly 80,000 a year. That's a twenty-four-hour workday, three shifts, and roughly a third of the workforce lives exclusively in the dorms. The rest commute, and not in minibuses, though the company provides free food, free housing, and free transportation to its workers, or at least those with "technical" qualifications.

This Brilliant knockoff costs somewhere between 60,000 and 120,000 yuan ($10,000 to $20,000), another step toward a future world of 2 billion cars. Beijing alone already has 4 million cars on its roads and adds at least 500,000 per year. Regulations permit only some of them to drive on any given day, but there is still gridlock, and the pollution enshrouding the city and driving its residents to an earlier death is increasing. Yet wealthy Beijingers circumvent the rules by buying two cars: one with an even-numbered license plate, one with an odd, so they can drive every day. The Chinese version of the American dream is not so easily stopped.

At the same time, it is difficult to squeeze onto the crowded Beijing subways at rush hour. There is no personal space, and the kind of crowding that leads to a fight in New York City subways is normal there. Honorable men stand frozen next to women so as not to touch.

Meanwhile, the old industrial cities of Liaoning Province, like Benxi, still make steel, 20 to 30 million metric tons of it per year. How else to make cars or help the national bird of China to proliferate? How else to raise high-rise apartment complexes, or wind turbines, or supply the shipyards in Dalian meant to one day crush those around the vast bay known as the Yellow Sea in South Korea? How else to contribute to the overproduction despoiling the world? Liaoning is hilly country, with villages and even cities nestled in the dales, an undulating country of unvarying brown in winter, but for the occasional dusting of snow.

There are the ambitious yet unsustained attempts to build an eco-city: Huangbaiyu, Dongtan, Tianjin. Perhaps first among these equals was Dongtan—literally "East Beach"—an ecological satellite city of Shanghai that would rise from the marshes of the alluvial island of

Chongming, across the mouth of the Yangtze from Shanghai's Pudong district. And although a bridge and a tunnel connect Chongming to Shanghai, no city yet troubles the marshes, which is perhaps good for the birds that use those marshes as one of the few remaining landing spaces in all of China. This city in a marsh promised to produce little trash thanks to its circular economy, relying on the plentiful sea breezes for power, and to permit no ordinary cars within its limits, only those that spew no CO_2 from their tailpipes, like hydrogen or electric cars. Hundreds of pages of plans, maps, charts, and even a glossy coffee table book laid out in great detail the inner workings of this ecotopia: energy-efficient buildings clustered close together to promote walking among the fifty residents per acre, organic farms arrayed around the new development for food, and a power plant fueled by rice husks that would also heat homes. Even the birds would find a comfortable home among the canals and ponds that brought the surrounding wetlands into the city's design itself.

Meant to be completed by 2010 when Shanghai hosted the World Expo, Dongtan instead remains a dream and a marsh, done in perhaps by the corruption arrests of its local government and developer champions or perhaps by the impracticality of its fantasies. Dongtan is a Potemkin village that never even got built.

Of the great Chinese ecocity projects of the early twenty-first century, only one remains, and it is the one that is taking place within an existing city—Tianjin—which also happens to be close to Beijing. Yet, Tianjin residents boast higher carbon dioxide pollution per person than residents of New York, London, and Tokyo. And it is existing cities where the future is laid out and experiments conducted. These are the *shengtai wenming jianshe shidian*, ecological civilization trial sites.

Rizhao is such a trial site. The seaside city with its enclosed marina is essentially a tourist community. Last year 27 million Chinese, primarily from inland provinces, visited to see the sea. But no one is accounting for all that carbon dioxide the tourists from Anhui, Henan, Hubei, and Shaanxi emitted getting to Rizhao, which was once more simply named: Haiqu. Go to the sea.

Still, all those tourists mean the example of Rizhao could spread. Perhaps the Anthropocene should be renamed in Mandarin, or Sanskrit. The developing man's burden will be nothing less than saving the world.

*   *   *

In 2004, the city that is first to get sunshine got its first use of said sunshine: hot water. The Shan Shui hotel (Mountain Water) installed solar hot-water heaters from Tsinghua University on its roof. Glass tubes with black carbon inside absorbed the sun's heat and transferred it to the water.

Now such solar hot-water heaters are the law, and regiments of them line the rooftops of every new building. Natural gas is cheap but sunshine is free, even if the systems cost at least 4,000 yuan ($600). Plus, Rizhao gets a fair share of it: 260 days of sunshine, according to Fan and the local weather bureau. "This is why we install solar energy," he explains. Still, on the day of my visit to Mountain Water hotel, it is raining in the sunshine city.

Shan Shui's owners are not the only ones to think this way. The White House agrees, twice over. In the 1970s, President Jimmy Carter installed thirty-two solar hot-water panels on the White House and predicted that "a generation from now, this solar heater can either be a curiosity, a museum piece, an example of a road not taken, or it can be just a small part of one of the greatest and most exciting adventures ever undertaken by the American people." The American people opted for the former via leaders like Ronald Reagan; the solar panels came down during a refurbishment in 1986. Subsequently rescued from government warehouse oblivion by a young administrator from Union College in Maine, they heated water for that school's cafeteria until they did indeed become museum pieces—one for the Smithsonian and one for the Solar Science and Technology Museum in Dezhou. After all, China now produces 80 percent of all the solar hot-water heaters in the world.

The White House is trying again. First, the George W. Bush administration added solar panels to some outbuildings, to help heat a swimming pool among other amenities. Then the Obama administration, as part of an overall refurbishment to help the White House waste less energy, added to the main roof the silicon panels that can turn incoming sunlight into electricity.

Sunshine to electricity might prove key to a world that burns less coal. China is also the home of much of the world's photovoltaic production. More than 500 companies compete to churn out solar technology, and the country has already seen the rise and fall of at least one Sun King.

Even as solar gets cheaper, however, it is still too expensive for most, and what is too expensive for relatively rich Americans is far out of reach for almost all Chinese. "If you want to generate solar energy, the costs are very high," Fan explains, and he's right. "But in the future we can use this technology." That future can be seen on some of the lampposts that light up the beachside at night.

Like the imaginary Dongtan and all-too-real pollution of Tianjin, Rizhao has a plan: The ecological city construction plan authored by members of the Chinese Academy of Science and Technology for Development. It is not just clean energy or the circular economy. A World Bank loan enabled the city to build a world-class sewage treatment facility, chlorination and reverse osmosis intended to make the water discharged into the sea clean, or at least cleaner.

Rizhao is already famous for its seafood, which is even exported to South Korea and Japan despite the pollution that still prompts algae blooms to coat the coast of the Yellow Sea. Those relatively new opportunities have made some rich: A big red archway covered in large Chinese characters promotes a wedding, an extravagance in the celebration of only children. A poster-size picture of the lovely bride and handsome groom is cut out in the shape of a heart.

Rizhao is torn between tourism and industry, especially the port. The most polluting industries have been shifted from the city itself to the surrounding countryside. They impinge on the villages but allow Rizhao to inch closer to carbon neutral. In any larger scheme, however, there is no away to throw away to, and the pollution that spews from factories on the outskirts of town has the same impact on the atmosphere as when the factory was in town. In some sense, climate change is the new front in a war between local governments focused on growing the local economy and the central government focused on preserving social stability at any cost.

Fan is not like other officials, according to the affable young man who accompanies me on my second visit to Rizhao, whom I will call Frank. (Frank is his adopted English name, which he chose because he wants to be frank.) Twenty-eight years old, he shelters his broad face under a baseball cap. He has a big, easy laugh that reveals the white teeth that brand him as a nonsmoker, a relative rarity among Chinese men. He complains of being too fat. Trained in science, he now works on history and litera-

ture at a small press, simply because there are too few jobs for physicists. He is an ardent believer in free markets and Austrian economics as well as being a Communist Party member, voicing near constant criticism of his own party. Such is modern China. "If you think you have big government in the U.S., come here," Frank is fond of saying.

There are a host of ways to get by in China as a government official, sinecures that involve tea drinking and reading the newspaper all day while living off of other people's taxes, taxes which officially do not exist except for those which draw from those who earn more than 3,500 yuan ($530) per month. To begin to get at the truth of all this, Frank would like to see a version of *The Daily Show* come to China. In fact, he thinks the recent change in policy to deemphasize teaching English is a plot. "I am totally against this trend," he tells me. "You must know English well to know your own country. Do you know what I mean?"

Fan knows English and is interested in chess and tennis—both the kind played on a court and on a table. He has a passing familiarity with American politics, the structure of the U.S. government, and even its historical roots, which seems almost subversive.

Fan likes to sing when drunk, as I discover when we banquet together at the locally famous Eight Bowls restaurant near the beachfront: just me, Fan, and an official translator by the name of Yan Hui. Fan has a favorite song in English, "Country Roads," and he wants me to sing it. I disappoint him by failing to know all the words. Or rather fail to recall them after numerous toasts of *bai jiu* that require me to drain my little cup of liquor. John Denver was a part of my youth that I thought I had left behind in the 1970s.

Yan admits she is quite fond of Americans, or at least has been raised to be so. Perhaps she is just being polite. At one point in the evening, to demonstrate the friendship ties between China and the United States, she brings up the American pilots who flew supplies over the Himalayas to the Chinese during the long war against Japan. That is a piece of history not often remembered on my side of the Pacific.

Eight Bowls' décor consists of the residue of fishing: nets, plastic crabs, portholes. What singing I do, horrible and off-key, I thankfully sing in private, in a wood-paneled room served by our own personal waitress. The three of us order twenty-odd dishes, an extravagance occasioned by the visit of this big-stomached foreigner. There is fried fish in pancakes;

shrimp and fish soup in a spicy, milky broth; some kind of flower bud for the vegetable dish; and then the ubiquitous dumplings for when one feels already too full.

The foundation of modern society is waste, literally. Battery Park City in New York rose out of the waters on dumped landfill from the building of the Twin Towers. This new land became the southernmost edge of Manhattan until Hurricane Sandy put it back in its place—underwater. Fully 20 percent of the land covered by New York City these days is the result of dumped fill, some of it garbage. The Gowanus in Brooklyn, where I live, is a filled-in swamp, remnant waters of a meandering creek and marsh channeled into a fetid canal. As it goes in New York, it goes in all other cities of the world. The older the city, the more strata of trash lie beneath it. Such past often brings the present to a halt, as a king's corpse is found beneath a parking lot or a new subway tunnel is halted temporarily to tease out ancient history.

Waste is the preeminent problem of the city, how to dispose of it—and it is a growing problem. Cities now produce as much as 10 billion metric tons of waste each year. Even the first city dwellers, a Paleolithic people known to archaeologists as the Natufians, precipitated the first garbage (and sanitation) crisis almost immediately upon settlement. After all, their nomadic ancestors didn't need to worry about trash or waste, moving from site to site and leaving their discards behind for scavenging and breakdown. The Natufians kept those ways initially. Their early settlements in the Middle East are strewn with garbage—and thus perhaps aided the creation of disease. There was a "garbage crisis in prehistory," as some archaeologists put it, and a garbage strike can still bring down a mayor in New York.

A New Yorker throws away an average of 1.2 kilograms of stuff every day, and a mountain rises in Staten Island from the garbage of the five boroughs. The grass-covered mounds that mark the Fresh Kills landfill stretch for kilometers in a former swamp, visible from space, an enduring marker of New York City's prominence that is both human and geologic in nature. Just as the ashes of a thousand home fires became the landfill that gave New York a home for the baseball stadium for the Mets, the U.S. Open, and Flushing Meadows–Corona Park, the ashes of the Twin Tow-

ers and the garbage of decades will turn the mounds at Fresh Kills into a verdant garden.

The newest part of Rizhao is a former lagoon that now hosts luxury hotels, a marina, and a venue for the local bourgeoisie's weddings. In only four years in the early 2000s, Rizhao filled in 9 square kilometers of lagoon, reshaped it, built on it. That's more than one-quarter of Rizhao's entire sprawling extent.

That is far from the only beneficial reuse of waste in this would-be circular economy. The fly ash left over after the coal is burned in the local power plant becomes brick, cement, and even cinder blocks, used for the continuous apartment construction. Waste even becomes energy, burned directly in Honolulu, for instance, where landfill space and fuels are both at a premium.

Yet still there is more waste, on the land, in the water, in the air. Air pollution—dust and soot—rises off the North China plain like steam from a lake in winter. The Great Pall of China completely obscures the country, turning the sky a dull yellow or blue blur, depending on the type of pollution. As the joke goes: The haze is a deliberate plot to hide China from the spying eyes of American satellites. Too bad it cuts short the lives of the Chinese themselves and travels all the way across the Pacific Ocean to the United States, the same path as the tidal wave of imports whose manufacture partially created it.

Some of that haze is pollution from the thousands of coal-fired power plants and millions of cars that China has added in the last few decades of unparalleled economic growth. But some of it is older: dust blown off the Gobi and Taklamakan Deserts, great arid swaths advancing on cities like Beijing and held back only by one of the largest geoengineering projects on the planet, the Three-North Shelterbelt Forest Program. In 1978, China began planting trees to hold back the desert; the goal is 100 billion trees serving as a Great Green Wall across a 4500-kilometer belt of northern China. Its efficacy remains in dispute, but no one can dispute that the Chinese, when they do something, do it big.

To diminish the mountain of coal burned every day to power China, the country plans other mega-engineering projects: a hundred or more nuclear reactors, the largest wind farms, and the largest dam in the world. The apparatchiks who run China were often trained as engineers, so engineering solutions seem to appeal to them. In some ways, it is reminiscent of 1950s America and the can-do optimism of scientism, the idea that science can be applied to solve social problems. China is being run by the same triad of military, industry, and science—men, always men, of progress—who ran the United States in the halcyon days of "I like Ike." Hu Jintao trained as a hydroengineer, Wen Jiabao as an earth mover, Xi Jinping in chemistry. Scientism run amok also provided the dams that litter the western United States, not all of them necessary, a mistake China is currently repeating as local governments rush to secure the jobs and money available for such projects.

All these nuclear power plants, wind farms, and dams are the only hope China has to keep its promise to stop using the atmosphere as an open sewer. That may prove a faint hope. As Frank says of the massive build-out of nuclear power plants, "Anything built in China cannot be safe. Too much corruption." That goes for the dams too, and the wind turbines often spinning to no purpose. At Daya Bay in the south, China's first nuclear reactor, a viewing platform shows where the steel bars necessary to reinforce the containment vessel were left out, forcing a rethink and modifications to buttress the reactor vessel in some other way. The concrete pier stands as a testimony to transparency, perhaps, but also to the kind of corruption that allows a construction firm to undercount the number of steel rods required, or to claim so in a time- and money-saving step. The very first reactor the Chinese ever built, at Qinshan in Zhejiang Province, had to be torn down and rebuilt because of flaws in the foundation and defects in the welding. And the former head of China National Nuclear Corporation has been sentenced to life in prison for corruption.

Heedless growth leaves a long-lasting mark, like the Superfund site where I live: the Gowanus Canal. This onetime meandering creek became an industrial canal, along which squatted the big factories to turn coal into town gas for lighting. The sacrifice for that light was tons of toxic tar left over after squeezing coal into town gas, tar that was simply dumped into the open sewer that is the local waterway. Gowanus became the sacrifice zone for the lighting of Brooklyn.

The canal also happens to be at the bottom of a large hill bearing many houses, conveniently situated for fourteen major sewer outflows, adding a layer of raw sewage on top of the toxic muck. When it rains in New York, that raw sewage continues to accumulate to this day.

Sewage is the evil twin of garbage, posing similar problems of "away." This evil twin is currently vanquished with the copious addition of freshwater, large vats where microbes digest humanity's leavings, and, when necessary, the old standby of dumping it in the nearest large body of water.

By forgetting history, mistakes are repeated, as philosopher George Santayana warned us. China is repeating the coal-to-gas mistake, given that the country has coal and wants gas to cut down on all that choking smog and soot. In addition to a problematic tarry residue, turning coal to gas has the downside of increased carbon dioxide emissions, even worse than just burning the coal directly. This is not the only repeated mistake. China's chemical industry turns rivers all the colors of the rainbow, and a flotilla of dead pigs is not an uncommon sight in the waterways. The rivers run black sometimes thanks to such effluents, says Ma Jun, one of China's preeminent environmentalists. Most such polluting goes unpunished, and those who do draw the ire of the local, provincial, or central government pay fines that are a small fraction of the profits gained from the pollution in the first place. "It is increasingly difficult to find clean water," Ma told me.

Some 300 million people—nearly the entire population of the United States—still have no access to clean water, and two-thirds of China's booming cities do not have enough water of any kind, a condition the government calls "water-stressed." Rivers are dry mudflats with smaller rivulets carving a path through the muck or pooling in fetid dead ends. The central government is building a massive viaduct to siphon water from the Yangtze River and bring it north to Beijing and beyond, a megaproject called the South-North Water Diversion, which forced 300,000 people to move out of the way. Meanwhile, waterless urinals thank patrons in English for protecting water, which is "a scarce resource on Earth."

Still, some lessons have been learned. Shanghai Chemical Industry Park is building a swamp to deal with its wastewater, cleansing chemical contamination in a habitat built for marsh life, like frogs, not people. I sit for hours in Shanghai traffic to reach this small fake wetland on the outskirts of town. Trees are everywhere. There is toad song, there is birdsong. Dragonflies perch on the edge of marsh grasses as towering wind turbines spin slowly in the distance. Crabs scuttle beneath the wooden bridges, and tadpoles join fish for a swim in multiple ponds. The polluted water flows in at a rate of 22,000 cubic meters per day, spends a week meandering through the 30-hectare artificial swamp, making the algae bloom, and comes out cleansed, prevented from seeping into the ground by a plastic liner that pokes up here and there like a partially buried garbage bag.

On one side of this water park is a village, and the village creeps in too, like the guy cutting through the swamp's wooden paths on his bike and local farmers hoeing just over the wall. As everywhere else in China, there is a farm on any available piece of relatively flat land, which includes terraced mountainsides from Liaoning to Yunnan. Styrofoam, plastic wrappers, and discarded cans of Suntory beer litter the verdant pools of this custom-built swamp. On the other side is the China Petrochemical Research Institute, which contributes much of the fouled water. The rest comes from the litany of chemical companies in the area: Air Liquide, BASF, Bayer, CITIC, DuPont, Fluor, Hagemeyer, Huayi, Mitsui, Sinopec, 3M.

This swamp is the best that can be done at present, and it is certainly far better than the Gowanus Canal, even if both boast blossoming rainbows of oil atop the waters. This same story of water pollution can be seen from Rio de Janeiro to Mumbai, where the bay bears paths charted in oil rainbows, amid the ferries jostling with an Indian armada, Russian tugboats, and German supertankers.

The problem is not a lack of technology or a lack of money; it is a lack of motivation, not unlike the world's lack of will to combat climate change.

Driving in China is a precarious affair, people honking their way through intersections amid a chaos of turning trucks, zippy motorcycles, and tuktuks puttering along on three wheels. Along the roadside, electrically powered bicycles cruise silently. Giant buses slot through city streets and past overloaded trucks by mere centimeters. When the inevitable happens, a little crash between two cars in the old, cramped part of Rizhao, no one gets hurt, but a traffic cop makes sure to document the entire scene with cell phone pics, while talking on another cell phone pressed to his ear. A crowd gathers to discuss the finer points of who is at fault, though I imagine it is hard to truly assign blame given the surfeit of new drivers and me-first ethos of driving in China. Frank himself, despite his desire for personal probity, had no choice but to buy his driver's license without ever touching a steering wheel. There will be plenty of time to practice on the road itself.

Drivers do whatever they want: reverse to make a missed turn, flip around in the opposite direction in the middle of a busy road, speed down the highway's edges—even strike and kill unwary pedestrians. Right of way goes to the most aggressive, and everyone relies on everyone else to adapt. Drivers don't use their windshield wipers when it's raining, instead turning them on once and then letting visibility decline to the bare minimum before flipping them on again, as if preserving the wiper blades for some future emergency. A left-turn signal and a flash of the lights means, "I would like to pass you." If that doesn't work, there's always the breakdown lane, the fourth, outside lane on the modern Chinese highway. Fortunately, nearly every road is new, which makes the driving marginally easier, except that within two years of any highway being built it is chock-full of traffic.

Road signs read BON VOYAGE, HAVE A PLEASANT JOURNEY, DO NO DRIVING WHEN TIRED, PLEASE FASTEN YOUR SEAT BELT, CONTROL YOUR SPEED, KEEP YOUR SPACE, OVERSPEEDING PROHIBITION, and there are even cutouts of fake policemen in a bid to trick drivers into slowing down. Rule by slogan is the Chinese way. Meanwhile, a convoy of high officials zips by at top speed with a full police car escort, a sight that has become more rare in Xi's China.

The air is dirtier thanks to all these cars. Beijing's highways have become sinuous car parks that inch along slower than the conveyor belts in factories, each car bearing workers toward office towers or industrial parks at the edge of the city. There is plenty of time to admire what can be seen of the scenery, including a bridge that bears famous equations: $E = mc^2$ and Newton's Law of Universal Gravitation, $F = Gm_2 x m_1 / r^2$.

In Shenyang in the far northeast, the traffic never stops, day or night, the Chinese rushing to the future just as heedlessly as Americans. Frank thinks all these cars are just another sign of conspicuous consumption, and he extols the virtues of electric bikes—bicycles with an assist from an electric motor—for "ordinary people." The roads of Rizhao are a mix of such e-bikes and Audis. The only flaw in the e-bike plan is that Frank has to lug the heavy battery pack up and down the stairs to his apartment every morning and night to prevent its theft.

Jinan, the capital of Shandong Province, has bus rapid transit and a fleet of alternatively fueled motor coaches, old hybrid electric buses connected to cables overhead that disconnect in a startling shower of sparks or buses that somehow burn urea and oil. Like buses everywhere, they are crowded, offering a cross section of society from the elderly to students in matching track suits obliviously minding their cell phones. The springs that make Jinan famous also prevent the city from building subway lines. No one is interested in damaging the clean water for public transportation, yet. So traffic is bad enough that a taxi driver can read the sports pages, replete with tales of Chinese basketball and European soccer, while ferrying a passenger across town. As a result, by some measures, Jinan has the worst air quality in the world.

The Chinese have built much more than just highways. Over the last decade, a network of high-speed trains has crisscrossed the country. Businessmen ply their trade between cities via these trains (and planes and automobiles), cell phones ringing with military marches. Out the window, the trains speed past more chewed-over mountains, as if some giant has been gnawing on them. That giant is the ceaseless demand for cement for new construction, to mix with the 13 billion metric tons of shattered rocks known as aggregate that humanity digs up each year. Soon the mountains themselves will be as flat as the plains, while skyscrapers will provide the new scenery.

The sleek trains have bulbous noses that help cut air resistance as they race along elevated tracks with a gentle rush and *throom*, like the old chugga chugga sped up and smoothed out. The countryside seems to exude a thin white haze of smoke and smog, the brown fields covered in crop stubble. The seats are narrow. The Chinese have not caught up to the Americans in girth. In that respect, at least, the Chinese dream is more narrow.

The train reaches a top speed of 302 kilometers per hour on its raised tracks, gliding above farm fields. A promotional video shows the life of a high-speed train engineer, looking out his bubble window as poles race by. He drives as if in a car, albeit at much higher speed. The video extols the new lines and tunnels built in just the last few years, big infrastructure propaganda. Barring catastrophe brought on by the corruption during construction, this infrastructure sets China up for a century of internal trade and, thus, wealth. Grid linesmen and track workers receive the same glowing video treatment, scurrying across high-voltage wires and tightening bolts to prevent derailments.

The bus is very different. "You just want it to be convenient and cheap," says Frank, noting that speed is not the main concern for buses, which does not stop drivers from employing the breakdown lane to pass.

On my bus trip back to Rizhao with Frank, we pass Mount Tai, one of the most sacred mountains in all of China. Smog renders it nearly invisible, even though this is technically a "blue sky day" for the area around Jinan. Tiny peasant plots cover the countryside, interspersed with the occasional village, some with the fancy doors on homes that denote new wealth and some still poor. Frank's duties for the Communist Party include working in one such poor village, which would really like a sewer system, an infrastructure improvement project for which Frank has no funds. Without funds, poverty alleviation is all window dressing.

On arrival, Frank and I share a taxi from Rizhao's bus depot to our hotel that runs on a tank of compressed natural gas so big there almost isn't enough room for luggage. Fueling up requires driving past a line of waiting cars and backing into a refueling station trunk first. On a per-kilometer basis, natural gas is cheaper than oil in China, even though the gas comes all the way from central Asia or Russia. The United States may talk about an all-of-the-above strategy on energy, but the Chinese really live it.

Everywhere there are more and more cars, growing by 20 or even 30 percent per year in some cities, including Rizhao, precluding any hoped-for carbon neutrality or even smog reduction. Frank thinks carbon neutral might be possible in some wealthy Scandinavian country like Denmark, but not in China. There is simply too much growing to do.

I meet Fan at another seafood restaurant. In some ways he seems even younger a few years later with his buzzed hair and skinny physique. The smoking helps keep him thin, perhaps, now that he no longer bikes to work. He claims to have gained 5 kilograms, but I cannot see it. While we eat, the ubiquitous television broadcasts the doings of Xi Jinping and the central government, as they war on corruption, unrest, and pollution.

Fan is now working on "total pollution control," though that totality apparently does not include carbon dioxide. He admits that there is no way even to restrain any kind of pollution during the present era of growth and industrialization. Maybe someday, he says, exhaling a cloud of smoke from yet another cigarette, but not now.

So many cleanup efforts have fallen by the wayside in China over the early years of the twenty-first century: green gross domestic product, green growth and, now, perhaps, carbon neutral. These days the talk is of how growth will be sacrificed for "clean development," but there is little sign of clean air and plenty of new coal-fired power plants, port expansions, construction cranes, and goods from cars to solar panels churned out by mass-production lines. Just as Frank cannot get the money for sewers in his village work, Fan cannot get the money for pollution-control technology. Factories cannot afford to install their own pollution controls and, as everywhere, environmental protection is resisted or, as Fan puts it, not all the owners are cooperative. They want to "play sneaky," he says, though emissions-monitoring technology and new rules including fines may help with that. And factories are much easier to handle than, say, the volatile organic compounds released at the city's growing number of gas stations.

The next day dawns peach and clear over the East Sea. Fan comes to visit Frank and me at our hotel for another friendly chat. Fan touches his ear as he tries to remember the past or some arcana from his pollution-control battles. He leans back in his chair and remembers a time when the tallest building in town was only five stories. Now several reach forty stories or more thanks to investors from other provinces. My taxi driver thinks all the new construction is just a hiding place for ill-gotten gains from mining. Miners get cancer while owners get rich. So the coal afflicts everyone from village families that lose fathers to Beijing residents breathing fouled air that cuts short their lives.

Villagers from the surrounding countryside are ushered into Rizhao, now swollen to 3 million people. And yet tall apartment buildings prevent the intensive use of solar hot-water heaters, because the available roof area is just too small for all those apartments. People leave for work before the sun has even risen, too early for the solar heaters to have warmed up their bathwater.

In Fan's youth, the air was "much better than now." Yes, Rizhao has always had marine haze, but it was different from the gray-tinged smog of today. Nevertheless, he cracks a window "for pollution"—in other words, so he can smoke. My friend Fan no longer cares to discuss carbon neutral plans, or even climate change, really. Instead, he emphasizes what improvements there have been on his watch in air and water pollution.

Fan always looks slightly concerned when talking, perhaps eager to get his point across, with one leg tightly crossed over the other. He gestures with both hands for emphasis, though the concern may truly spring from speaking to a foreigner, and a journalist, no less. Frank's assessment of Fan is an honest one: He has achieved the "golden bowl," with the perks of a civil servant as well as a wife making money in the private sector. His daughter was even allowed to tour the United States for a few weeks, though this inspired in her a longing to go to Harvard, which is "not possible," according to her father. Still, the influence known as *guanxi*, as well as money, "is not a problem for them," Frank says.

That afternoon, Fan comes back with a driver in his special "0 0" government license plate car to give me a VIP tour of the port, the seashore, and other points of local interest. The first stop is to show how tall buildings can still employ solar hot-water heaters—by hanging them off their balconies. There are bigger yachts in the marina, and any algal blooms have been tamped down by the crisp fall weather. "Better than Qingdao," Frank says. Another bride wearing a brilliant red wedding dress has her picture taken in front of the glamorous boats.

On the other side of town, a Great (Plastic Netting) Wall has been built around the entire port at great expense to keep iron ore and coal dust from blowing into the city. Water trucks prowl the streets, spraying to keep down any dust that does escape. To cut down even further on dust, trucks carrying dirt have netting on top, something that never happens in other cities, according to Frank. There are even a few trucks powered by liquefied natural gas to decrease the soot from diesel burning, a problem that plagues ports the world over. A giant conveyor belt is being built to move iron ore and coal from the port and eliminate 20,000 truck trips per day, according to Fan.

To get to dinner after the tour, Frank, Fan, and I stroll through Silver River Park. The Silver River is the Chinese name for the Milky Way, our home galaxy. The park is really an old quarry transformed into a pond and greens dimly lit by solar lanterns. The real Milky Way is invisible, but I do actually spot a star in the sky, a first for me in urban China.

Rizhao is different from the gray-tinged hellscapes of, say, rare-earths-rich Baotou and smog-socked Beijing. But why?

I fly back to China on the first shopping holiday of the country's own invention. November 11 (11/11) has become known as Singles Day, just look at all those 1s in the date. But it is really another orgy of consumption, like Christmas in the United States.

The current ideology of the Communist Party of China is to turn the country from the world's largest factory into the world's largest market. If 1.3 billion people begin to consume as Americans, Europeans, Koreans, and Japanese consume, what is already the world's largest economy could grow exponentially. Already, malls that would not be out of place in the tony suburbs of Dallas or Durban can be found in every city, the only difference being some of the shops and wares on sale as well as the preponderance of Chinese faces. Stores hawk wildly expensive Western brands, most manufactured here in China, and there is even a food court, flavoring the air with spices.

Flying into Beijing in the evening, I see the city lit up like a field of stars. It is a complicated calligraphy of light spread out across the tableland, like a constellation. Dark night is a scarce commodity these days in the richer parts of the world, and scarcity gives value. I'm not even sure I use the rod receptors in my eyes anymore, living in a city that never shuts off the lights. But this missing of the dark is a luxury and one that many Chinese living today can remember going without. Astronomers and night sky associations clamor for darkness, but the poor would like to escape the fickle, dangerous, and expensive glow provided by kerosene.

China will never light its cities all night like the wasteful United States, Zhang Guobao, the former vice minister of China's powerful National Development and Reform Commission, tells me. But he is wrong about that, even if some lights are turned off. It's not just in Beijing, as I see in Jinan when I rise before dawn to catch a train. This provincial capital has sprawled to cover 20 kilometers or so from east to west between the Yellow River to the north and the mountains to the south, and it is lit through the night. And it's not just streetlights. There's also the light show from the stadium nearby that runs all night with few, if any, spectators. The stadium doesn't even have a home team anymore.

Vice minister is not just a title, it is a way of life, despite President Xi's anticorruption campaign. White-gloved young ladies, or *xiao jie*, serve tea to these older gentlemen with shining black hair, freshly pomaded like Ronald Reagan's. The more powerful the man, the more comely is the *xiao jie* serving his tea. Meetings take place in imperial boardrooms outfitted with gleaming wood and marble, large murals of mountains with calligraphy, and vases from all the schools of Chinese pottery. The more powerful the leader, the bigger and more resplendent the vase, like the stunning examples stored in the palatial rooms where former president Hu Jintao held court in the Bird's Nest stadium during the 2008 Summer Olympics. I head upstairs at the NDRC, the central planning body of the Communist Party's central government in Beijing, and into a great hall tiled in marble, footsteps ringing as *xiao jie* scurry before me. A massive Ming vase stands like a guard outside the main meeting room of the vice minister. It is a working life of luxury and the grandeur that speaks of imperial power.

Each dignitary is flanked by lesser dignitaries in turn backed by an array of unsmiling factotums, bureaucrats arrayed to provide any information possibly necessary to the one speaker, the highest-ranking person in the room. Only the most disheveled of these bureaucrats does anything, producing facts and figures on demand for the vice minister when teased with a promotion. These men spend most of their days finalizing five-year plans or ensuring that the targets laid out therein are met, even if meeting an energy efficiency metric means shutting down coal-fired power plants at the end of a given month and blacking out a city. That, in turn, leads to diesel shortages as savvy businesses fire up generators to tide them over. But at least the government has an energy plan.

Those people born in the latter half of the twentieth century have seen China undergo a great transformation, from famine to a steering force of global capitalism. It is a transformation the world may never see again, given the messier realities of an India or a Nigeria. Everything has been changed; the neighborhood where one was born has been leveled for new office buildings, hotels, high-rise apartment blocks, or maybe even an Olympic stadium like the Water Cube.

Beijing is still burning, though. The air still stings the nose and reeks of smoke. The trees that line the road from the airport are taller these days, blocking the view, just as the Chinese government blocks the view of the *New York Times* and Twitter. The city sprawls yet farther into the countryside, undertaking construction of a sixth ring road that, in places, is hundreds of kilometers from the city center.

Even the ecocity in Tianjin cannot compete with such sprawl in the villages on its outskirts, which reach almost to the edge of nearby Beijing. Random apartment buildings rise up out of agricultural fields to no discernible purpose other than official statistics. There is a property bubble in China, which is an all too real waste of resources. All of this registers as economic growth, whether it's useful or not.

The edicts of any government are only as good as those following them. Take the war on pollution. Rizhao's coal-fired power plant has pollution-control technology for two of its four units. Controls for the sulfur dioxide that causes acid rain are slowly being added. The six local kilns that churn out cement have technology that cuts down on the smog-forming nitrogen oxides that would otherwise go up the smokestack. Small boilers are being shut down throughout the province in favor of big, efficient ones like those at the big power plant. Fan describes all this as "very tiring work."

Even with the controls, however, the question remains whether the technology even gets turned on. That decision has more to do with the powerful power company Huaneng and its relationship with the central government than anything the municipal government might command. Plus, the power plant must provide the city's heat in winter despite all those solar hot-water heaters.

There is no immediate prospect for Rizhao or even Shandong Province to switch away from burning coal to burning natural gas to make electricity, even if fracking comes to China in Sichuan Province or elsewhere in the west. Imported gas from Russia is simply too expensive. And there is not enough marsh gas to light a city.

And then there is the fact that even with such controls at big industrial sources, the burgeoning fleet of cars makes the point moot. So although the amount of pollution from Rizhao's factories, power plants, and port has gone down thanks to Fan and his colleagues' efforts, the "air quality [is] a little worse," Fan admits. The "main problem may be vehicles." Such cars are "impossible for Rizhao to control," Fan adds, taking another drag

on his seemingly endless procession of cigarettes. He has to get his smoking in while away from his wife and daughter. "People always are buying
new cars in China."

Religion may provide some hope for China, as a friend of mine in Beijing avers. A return to Confucian ways born on the banks of Shandong's
rivers or the Buddhism imported from far-away India, maybe even Dao-
ism. Already the religion of organic food has spread to the wealthy here.
I'm not sure that any of this will save China from heedless consumption
and the pursuit of convenience.

Some of the necessary work in China is simply to imagine a different
future, cities built better, happy people, life beyond the daily grind. The
Chinese people can certainly make the future they want to see. A key to
that will be seeing through illusions.

Rizhao is not actually the first city to get sunshine. Japan has it beat
and, by international rules, Apia, the capital of Samoa, is the true first
city in the world to get sunlight each day. But perhaps Rizhao can be the
first to show how sunshine can be better used in China. Meanwhile, the
megacities of Beijing, Guangzhou, and Zhenjiang have all pledged to stop
growth in CO_2 pollution by 2020—and the country as a whole plans to
do the same by 2030.

On my last night in Rizhao, Fan, Frank, and I have our own private
banquet, a feast among friends, as Fan puts it, though that may be a
defensive posture against any more of my talk of carbon neutrality. Talk,
lubricated by a local rice wine, covers hobbies, books, movies, and the
American cultural imperium currently sweeping China and, in particular, Fan's home, given his daughter's fascination with the United States.
For her father's part, he liked *Argo* and thought *Lincoln* "a good movie."
In Shandong, a feast means a few opening dishes strictly for taste and
pleasure, such as a seafood soup with tiny corn dumplings or moist fish
in a savory brown sauce, followed by dumplings or noodles for real sustenance. Shandong cuisine is notoriously salty, forcing me to drink more.
The two Chinese men struggle to keep their shot glasses of liqueur lower
than mine as we toast, a sign of respect. "We are friends, we are friends,"
Fan assures me, looking me in the eye to make sure I understand.

Shandong Province is now a pilot area for the circular economy at a
provincial level. And if old Shandong can become a circular economy,
then perhaps so too can China as a whole. The Chinese dream does not

have to be a nightmare for the rest of the world, one way a better Anthropocene might come to be.

Just as one man in Tiananmen held back the tanks with a single act of courage, single acts like the ceaseless work of Fan and his compatriots across the country can stand athwart the juggernaut that is the fossil fuel economy and begin to shift it, ever so slightly at first but then more and more and more, in a new direction. In 2014, for the first time in decades, the amount of coal China burned fell.

That can happen even if carbon neutral has been left behind, a failed policy experiment. Even the United Nations Environment Programme has abandoned its efforts to promote carbon neutral and aid places like Rizhao.

After the feast, Fan drops me at the hotel after extracting a promise that I will meet his daughter. Her English is flawless, thanks perhaps to all those movies, or the grueling thirteen-hour days of high school in China. Students are turned into machines for tests here. Her dream is to go to Harvard but the Test of English as a Foreign Language, the SAT, and money stand in her way. Maybe for graduate study, as she is a scholar like her father and as gracious, welcoming, and friendly as her mother. I think she will make it. She has already enrolled to study economics at New York University's campus in Shanghai, the first major city in China to ditch economic growth targets.

Chapter 7

The Long Thaw

It was pleasant to believe, for example, that much of Nature was forever beyond
the tampering reach of man—he might level the forests and dam the streams,
but the clouds and the rain and the wind were God's.
—RACHEL CARSON, *ALWAYS, RACHEL*

There is a mountain in Oman that holds a billion metric tons of carbon dioxide, the odorless, colorless gas invisibly piling up in the atmosphere and trapping more and more and more of the sun's heat to change the climate. A lot of mountains hold carbon, but this one is special. It gets CO_2 fresh from the sky with the rain. The geologist Peter Kelemen viewed that as an impediment when he first set eyes on the peak in the Hajar range, back in 1994. Veins of carbonate, like skeins of white thread running through the rock, obscured what he was really there to see: Earth's hidden mantle, pushed up to the surface 90 million years ago by the movement of the fifteen plates that make up Earth's crust.

The mantle is the semisolid layer between the planet's core and the outer shell that floats along as the surface of Earth. Geologists sometimes describe this molten layer as plastic, and it oozes to the surface where oceanic crust splits deep beneath the Atlantic. Studying the mantle is thus extremely difficult except in those rare places where the crust's drifting has pushed mantle up to the surface and trapped it there, as in Newfoundland and Oman. So the barcode-like series of dikes written in the Omani rock informs our current understanding of what's going on with roughly two-thirds of the planet's rocky surface, the parts hidden by water.

In Oman, rocks that have not touched the atmosphere in many long years of geologic time now lie exposed, naked to the effects of wind and water. So the bearded Kelemen, a deadpan scientist and former mountain climber as solid as the rock he studies, makes an annual pilgrimage with geologists, biologists, students, and other fellow travelers along the rim of a steep canyon in the Hajar mountains, following the road up from the shores of the Arabian Sea. The caravan stops at a high point, follows a treacherous trail down to the valley below, and then into a fissure-like cave. The cave leads to a path that skirts the edge of a steep cliff, requiring ropes to prevent catastrophe, which then leads to a bridge beyond which is more trail that winds yet farther down to the bottom of the valley, where all is shadow.

There the abandoned village of Wadi Fins sits, its oasis around a big pond still intact and surrounded by the steep walls of the canyon. Those same walls help blunt the impact of the sun, shading the area for entire days in winter. That makes the work of studying the rocks or sampling for the microbes that live far underground much more pleasant than in the harsh light of the Omani desert, where summertime temperatures can reach 50°C.

The uppermost part of the mantle is primarily composed, geologists think, of a greenish rock they call peridotite, named after the gemstone peridot, which is pale green like a fresh spring leaf encased in ice. Kelemen hoped to study the chemistry of this rock to reveal something about the chemistry in the inaccessible mantle, more than 15 kilometers deep in most parts of the planet. Peridotite is an unusual rock, mostly made up of the minerals olivine—a mixture of magnesium, iron, and silicon bonded with oxygen—and pyroxene, a similar compound that swaps out the magnesium and iron for other common elements such as calcium and aluminum.

The Omani peridotite hid its secrets. Because wherever the rock was exposed to air and interacted with water, it sucked in CO_2, making a limestone-like rock. Rain falls on these rocks, dissolving the calcium inside, as the water seeps underground. When this water percolates back to the surface with a basic pH like soap, it greedily sucks CO_2 from the air to form a scum of calcite, dubbed travertine by geologists once it solidifies into rock. This chalky dreck made it difficult for Kelemen to find a pure sample of the mantle rocks he had come to study in the 1990s.

It wasn't until 2007 that it occurred to him that all this chalky rock could be the answer to a crisis bedeviling the whole globe, rather than his own personal problem. Instead of looking at the chalky white veins as obscuring the evidence, he realized he could see exactly what the talc revealed about how rocks help control the levels of carbon dioxide in the atmosphere. Given that people are putting as much CO_2 into the atmosphere as possible these days, burning fossil fuels, clearing forests, herding cows and their methane belches, the question became: How much CO_2 exactly did this mantle rock suck up?

Kelemen is skeptical of the computer models that aim to project the course of climate change. That does not, however, make him skeptical of the basic fact of climate change: CO_2 traps heat. "Even if you're conservative, if we continue up the exponential energy growth path and we use fossil fuels, we're going to see up to three degrees [Celsius] warming by the end of the century," he predicts. As it stands, pollution from people has already overriden the effect of wobbles and perturbations in Earth's orbit of the sun, forestalling the next millennia-scale advance of glacial ice.

If people want to reduce that heat, remove some CO_2 from the air, like Smetacek's dream of fertilizing the oceans to induce a carbon-sucking plankton bloom. Another key to that might be those minerals in the peridotite, Kelemen thought: magnesium and calcium that wanted to react with all that CO_2 in the air to form this talc, a talc that could be stable for millions of years. It might be the perfect rocky prison for the excess carbon formerly stored in fossil fuels. The problem: Rocks, like plate tectonics, work slowly. Or so everyone thought.

SPEAK TO THE EARTH AND IT SHALL TEACH THEE reads the motto on the outside of the building where Kelemen shares an office at Columbia University. This is Schermerhorn Hall, named after the son of one of the Dutch shipping and rope-making brothers who also gave their family name to a street in Brooklyn and thus, ultimately, a subway stop, their very own bit of anthroturbation. The building is made from the products of geology: rock and concrete. Like the subway builders of yore, Kelemen would like to drill too, only in Oman. Because the weathering of the peridotite may be the key to solving global warming.

Such weathering is how Earth recycles CO_2 on the million-year timescales of geology. Or faster: "I could throw a pebble in one of those pools and it'd be covered in calcite in a day or two," Kelemen explains. "From a geologic perspective, that's supersonic."

Kelemen thought it could be even faster. Carbon dating revealed that the carbonate veins in the rock are young, 50,000 or so years old, or roughly as old as the oldest iteration of this newly proposed epoch and forming as the ancestors of all the humans outside Africa migrated through this region to populate the planet. Essentially, new peridotite would be exposed by erosion, the talc would form and, in turn, be eroded away over a few tens of thousands of years, allowing the process to repeat. In this way, the mountains in Oman suck up 100,000 metric tons of CO_2 per year. It is an ancient process, air and rock conspiring.

The capacity of these mountains and others like them around the world—including those at the bottom of the sea—is gigantic. Every gram of this rock can suck up 0.6 gram of CO_2. "Rocks are heavy," Kelemen notes. "You could probably take up 33 trillion tons of CO_2 in Oman alone if you could carbonate every single magnesium atom in those peridotites."

That's one 350-kilometer-long mountain range, maybe 15 kilometers wide and, assuming access to the top 3 kilometers' worth of rock, a chance to suck up 1,000 years' worth of present-day pollution.

Of course, that's not what's happening. The exposed rock in Oman takes out a minute fraction of the 36 billion metric tons that come from fossil fuel burning.

"Not a lot," Kelemen says with a tight smile. This is the same man who, when I congratulated him on being appointed academic chair of his department at Columbia, offered that condolences would be more appropriate. He winces when he talks about the challenges of chairmanship.

On the bright side, as Kelemen adds, "If this was happening too fast, we'd all die because there wouldn't be any CO_2 in the air. The rate is the issue." And the rate could be faster, but not too fast. Or so Kelemen thinks. The only way to find out is to drill. And nobody wants to let him do that.

To understand the sky, drill down. Ice cores wrested from Greenland and Antarctica hold bubbles of trapped prehistoric air. These are time capsules bearing messages from the past. Scientists measure the little puffs to find out what the atmosphere was like before humans. That ice record can travel back a million years or more, but it has nothing on rocks. Red bands in the desert tell of the time when iron undissolved out of the ocean waters thanks to rising levels of oxygen, scientists think.

There are also the thick bands of coal that tell of ancient swamps and trees, too numerous perhaps for fungi to break down the complex sugar molecules that held the foliage up higher than ever before. Since then, the atmosphere has fluctuated around mostly nitrogen, somewhere around 20 percent oxygen, and 1 percent or less CO_2, methane, and other tidbits. It is on the cycling of that 1 percent that life on Earth—and climate change—depends.

Drilling offers the possibility of finding the right rocks below and thus (it is hoped) the chance to change the atmosphere today. The problem is carbon, a minute fraction of a minute fraction of the atmosphere. Over geologic time, the carbon in the atmosphere has moved from sky to plant or shell, and then, once buried, to rock, which ever so slowly gets crushed and subducted in the endless march of plate tectonics. This crushing march drives the entombed CO_2 back underground, where heat and pressure liberate it to be spewed back into the atmosphere by a volcano. Or at least that's how it worked for billions of years before humans came along.

The first part of that cycle is the essence of modern life. The carbon floats in the atmosphere paired with the oxygen that could drive life—if freed from the grip of carbon. It is the patient carbon knitting done by plants with the help of a steady input of energy from the sun that makes this possible. Oxygen, that unstable ravager of an element responsible for all kinds of corrosion from rust to cellular damage, remains abundant only because of all the plants busy with their knitting on land and, even more important, at sea. All those cells taking in CO_2 and spitting back out O_2 contribute vast volumes of this oxygen to the atmosphere, enough to cover the entire surface of Earth in a layer 6 centimeters thick. Add to those efforts the Sisyphean work of their fellow photosynthesizers, and the planet ends up with an oxygen concentration of 21 percent, high enough to bury the planet in more than a meter of solid O_2 if it wasn't a gas.

To survive, all living things breathe in and out trillions of liters of air every day, ignoring the primary component of air, nitrogen, but greedily taking in the oxygen to fuel the breakdown of carbon chains that powers the building of yet other molecules necessary for life. The chemist Carl Wilhelm Scheele called oxygen "fire-air," and it is oxygen that makes fire, as well as human life, possible. Abundant microbial life is the champion of all this breathing, from the hidden deeps of the rocky earth all the way to the stratosphere, burning through life by breaking big molecules down into smaller molecules with the help of oxygen.

The minute proportion of CO_2 in the air is now rising, ever so slightly. It is not because life has become more abundant; the exhalations of even 7 billion people or the tens of billions of cows, pigs, and chickens amount to nothing compared to volcanoes. No, human pyromania bears the blame for that. The atmosphere provides the medium for fire, a flame without a memory that imitates animal life by breaking carbon chains, releasing energy to further fuel the flame and off-gassing CO_2. When a tree, a hunk of coal, or a liter of gasoline gets burned, some of that copious oxygen bonds with the carbon in the fuel and wafts off into the atmosphere as carbon dioxide.

This tiny proportion of CO_2 and other greenhouse gases, including water vapor, plays an outsized role in the atmosphere. The greenhouse gases, led by CO_2, help provide a warm blanket for the planet. The mix of atmospheric gases lets sunlight in, but CO_2 and the other greenhouse gases trap some of the heat that would otherwise flow back into space. Without that greenhouse effect—courtesy of the minute fraction of CO_2 and water vapor most visible as fluffy clouds—the planet would average a wintry temperature of $-18°C$, a nice round zero on the Fahrenheit scale.

Lately, this proportion has begun to grow and, like any greenhouse with suddenly thicker glass, the planet has started to warm, headed toward temperatures that may prove uncomfortable for its inhabitants. The human mania for burning has now changed the amount of CO_2 in the atmosphere by roughly 100 parts per million. Not much. But that 0.01 percent has proved enough, if Earth's geologic history is any guide, to shift climates from vast glaciers more than a kilometer thick burying the spot that currently hosts New York City to the relatively balmy temperatures of today that allow a global capital to flourish.

All that CO_2 in the atmosphere would make a layer only 3 millimeters thick if used to cover the surface of the planet. Yet changing the composition of the air by 0.01 percent captures roughly 1.5 watts of extra heat per square meter of Earth's surface. This is the warming in global warming. And that much warming is no trifle. It's the difference between an icy Arctic and a warm sea guarded by crocodile-like animals in the planetary history inscribed in the rocks.

Global warming is an old problem, even on the human timescale.

Carbon dioxide began its modern rise in the 1800s as Europe recovered from the Napoleonic Wars and the Qing dynasty that had ruled China for centuries stumbled toward collapse after fighting two Opium Wars against Great Britain. Throughout this turbulent period, nations, merchants, and capitalists increasingly turned their attention to manufacturing and trade, burning more and more coal to power factories and steamships, and clearing woodlands. The result was that CO_2 levels surpassed the highest levels seen in an ice record stretching back 800,000 years, somewhere around 1875 as the United States recovered from the Civil War and cut down the last of its great forests. Since then, CO_2 has been at levels in the atmosphere not seen since before any kind of *Homo* existed, reaching 310 parts per million by 1945 and surpassing 400 ppm today. Modern civilization continues to engage in this truly massive, globe-spanning effort to liberate yet more carbon, undoing eons of the plants' patient knitting to get at the fossilized sunshine. Today, as never before, the sky is menacing.

Civilization burns through this entombed sunshine a million times faster than plants have laid it down over eons. Each of my fellow Americans spews an average of ten cars' worth of carbon dioxide each year—20 metric tons of an invisible, odorless, colorless gas that disperses through the atmosphere, invisibly trapping heat. All of us collectively, the wealthy few more than the numerous poor, add nearly 40 billion metric tons of CO_2 to the atmosphere every year. Some of those molecules will likely still be in the atmosphere more than 1,000 years from now—unless something is done to take them out.

Atmospheric intervention is old hat for humanity. The first farmers added more methane to the air, not a lot, but perhaps enough to create the stable climate of the Holocene. More recently, civilization has begun mining the sky. In the late nineteenth century, men of science, and they were mostly white men with a strong sense of burden, noted the coming crisis in fertilizers. The deep deposits of bird and bat shit on the islands of the Pacific Ocean and in the caves of South America that had fed the swelling European population would soon be gone, or taken over by the people actually living in those regions. And this at a time when the call for explosives had never been higher, given the need to suppress rebellions and fight fellow Europeans, explosives that relied on the same nitrates as crops. This was an intolerable burden, and one that science should fix.

So, in what is perhaps one of the biggest instances of directed scientific research, a call went out to figure out how to get at the nitrogen just sitting there in the atmosphere. An intolerable 68 percent of the air went unused, except for what the masses of microbes could turn into nitrates.

The Germans got there first, thanks to the chemical genius of Fritz Haber and the industrial engineering of Carl Bosch. High temperatures and pressures created by burning fossil fuels allowed iron-based catalysts to pair with hydrogen wrested from natural gas and nitrogen from the air, creating ammonia, nitrates, and urea. Germany's defeat in World War I ensured that all nations soon had access to this technology and humanity was freed from microbial limits on the nitrogen needed to make the proteins that allow our bodies to live. As a result, the human population swelled from less than 2 billion in 1909, when Haber first demonstrated this process, to more than 7 billion today—and growing. There are downsides—dead zones in the oceans from nitrogen fertilizer washing off fields, and smog in the skies—but without this process, at least half the humans alive today would not exist, and the world's forests and grasslands are more lush as well. This is, by far, the largest human intervention in a planetary-scale biogeochemical cycle to date, putting us on a par with all those teeming microbes who have had eons to perfect the technique. Now we want to do something similar, but smaller, with carbon, particularly removing excesses of it from the atmosphere, just like the cyanobacteria and their descendants, the plants.

Carbon dioxide comes out of the atmosphere in three ways. There is the patient knitting of those plants, floating in the sea and towering over land, and even the living soil itself can bury carbon. But that gargantuan invisible effort is not even the first line of defense against excess CO_2. Earth is a blue planet, ruled by water. The oceans absorb more than half the extra CO_2 at present, mostly directly into those waters. And then finally there is the geologic way: weathering.

Ah-Hyung Park (also known as the Carbon Lady) says to call her Alissa, with a laugh that rings out like a melodious bell. Her goal is nothing less than to save the world, just like her Columbia University colleague Kelemen, and to that end she keeps busier than most and casts a gimlet eye on the work of her colleagues.

Park and Kelemen are only two of a band of would-be atmospheric saviors, many concentrated at Columbia University. The band is just as fractious as any other loose grouping of academics. Discipline is not the rule in interdisciplinary. As the saying goes, the fights are so fierce because the stakes are so small—except that in this case the stakes may be the fate of the sky and its ruling influence on civilization. So perhaps the usually good-natured competition is to be expected among Kelemen, the wild-haired physicist Peter Eisenberger, the inventive Park, and the tall, prim, and kindly physicist Klaus Lackner, who has now decamped to Arizona State University to actually try to build the giant artificial tree of his dreams. Then there are a slew of other Columbia geologists: Dave Goldberg and Paul Olsen, who prefer volcanic rocks in the Palisades, just across the Hudson River in New Jersey, or just offshore in the Atlantic, to any Omani peridotites, at least for capturing carbon.

To speed the cause, unite like minds, and air grievances, a conference was held one spring, an attempt to bring together everyone from all around the world working on the problem of removing CO_2 from the air, whether by capturing it in a smokestack or sucking it out of the sky with a device or with duckweed, an aquatic plant. Park wants to take CO_2 from the air and convert it into chemicals or fuels, turning a problem into a solution like Majumdar and his ARPA-E cronies. Kelemen wants to turn CO_2 into rock, as does Goldberg, just different kinds.

The chemists at MIT and Caltech want to harness photosynthesis and the dark art of catalysis to turn CO_2 into useful products—plastics or hydrocarbon fuels. Denmark has a similar national ambition. "They call it waste to work," Park tells me, with just a hint of her native Korean inflecting her English, followed by a trill of laughter. "I don't think it's work yet."

Park couches her criticisms in laughter, perhaps to soften the blow. Of harnessing biology to get rid of atmospheric CO_2, she can offer only, "That's cute." How about turning all that CO_2 into some kind of carbonate building material, perhaps to displace concrete or drywall? "With the amount of CO_2 we have to store, we're not going to turn everything into valuable material," she says with another laugh and a disappointed shrug.

Park is more impressed with the storklike Lackner, who dreams of right-sizing technology, noting that microwaves and computers continually improve, especially compared to gigantic nuclear reactors still stuck in 1950s iterations or even coal-fired boilers changed out only every few decades. He hopes that his artificial tree can follow the mobile phone path, with rapid iteration and improvement, though size matters when it comes to efficiently pulling CO_2 out of the air.

She laughs almost as much as she throws out intriguing ideas for investigation. "I come up with so many crazy ideas, my students hate me," she says, waving the ideas away with her hand as if they were a plague of flies. Her list of wonders wanders from chemicals made from CO_2 to waterless fracking of shale to get at the natural gas trapped inside. "I can't stop fracking, so what I can do is: Can we do it better?" She has found that certain shales cooked in supercritical CO_2 crumble. "Isn't that crazy?" she asks with a cock of her head at me. "I cannot come up with a new fracking process, I'm not that person. But I can tell you what happens to a geologic formation."

Fracking with CO_2 might also be a good way to keep the climate-changing gas underground. Current fracking relies on tiny bits of sand to hold open the cracks created by the high-pressure water. With the right little particles instead of sand, the particles could grab onto CO_2 chemically and hold it in place underground. This would reduce water use, trap CO_2, and keep dirty water from making it back to the surface laced with long-buried poisons, like radioactive elements. "Pretty cool, huh?" she asks. "Whether it's going to work or not, I don't know." She laughs again.

She grew up in South Korea as that country completed its transition from poverty to the prosperous future, enabling for the first time the luxury of worrying about the environment. She now sits surrounded by the gadgets of that affluence, many of them Korean or bearing components made in Korea: several computers, each with multiple screens; a few laptops; an iPhone. "Whatever I have, I want others to have it, but with less environmental impact," she explains, musing on all the coal burned just because people don't turn off computers or lights at night, something that would never have happened when she was growing up in Korea. "It didn't take a long time to deprogram me," she says with another laugh, and looks around at the abundance moving to America to become an academic has brought her.

The most insidious part of that pollution is invisible. The denizens of New York banded together to get the stinky, dangerous trash off the streets in the nineteenth century, but it is harder with CO_2, invisibly piling up in the atmosphere without any scent at all. Even the pictures of smokestacks billowing what looks like white smoke are just showing steam: You can't see the worst pollution, only the shimmer distortion of heat coming out of a smokestack.

So Park gathered this real mad-scientist conclave, all working on how to suck CO_2 back out of the sky and alter the composition of the air we all breathe. What could go wrong? The scientists assembled in the bowels of the Schapiro Center for Engineering and Physical Science Research, speakers arrayed on a dais and facing an amphitheater of a hundred or so very smart men and women, Asians, Europeans, and Americans, all setting themselves against one another and the largest environmental problem humanity has ever faced. Fortunately, there was ample coffee. The audience hurled barbed words at the speakers, which were returned with force in PowerPoint slides of formulas and bullet points.

The current front-runner among technologies to capture CO_2 is a bid to catch the millions of molecules before they waft out of the smokestacks at coal-fired power plants. Some form of burial for that ubiquitous by-product of flame is also required, and this tech fix goes by the initials CCS, carbon capture and storage. The problem with capturing CO_2 from air is that if doing so became cheap enough and good enough, we would no longer need to stop spewing CO_2 in the first place, Lackner points out. In other words, we'd be trapping carbon to free coal.

Eisenberger worked himself into a frenzy about all the billions of dollars wasted on developing CCS—because it will not solve the problem. "Why spend so much time and energy and ingenuity coming up with solutions that are not really solutions?" he asks of the small crowd of true believers, his fringe of wild hair shaking. "All that learning is wasted. It's a dead end."

But the proponents of CCS would not be cowed. "We don't really have the technology yet," argues the implacable chemist Søren Lyng Ebbehøj of the Technical University of Denmark. "There is a gap we need to bridge here." In other words, coal will be burned, and thus there is a need to at least prevent all that CO_2 from ending up in the atmosphere.

Eisenberger, in turn, dismisses all the arguments in support of cleaning up coal with several annoyed shakes of his head, calling out anew in his reedy voice roughened by smoking, "We would have never got to the moon unless we decided to go there," he argues. "The fossil fuel community is going to be very happy with avoided carbon." After all, once the CO_2 pollution is cleaned up, what is to stop more fossil fuel burning? "We are creating monsters that we'll have to address. Air capture is the only solution that addresses the fundamental problem."

Eisenberger has just such a solution available for sale. It's called Global Thermostat and can supposedly suck CO_2 right out of the air or from a smokestack. He even has a unit up and running in Silicon Valley, testing the premise.

"What about cement?" asks John Hansen of Danish catalyst company Haldor Topsøe. "There's no way of avoiding CO_2."

"Not true!" shouts out Eisenberger.

But it is true that cement making is responsible for a huge amount of CO_2, both in the burning of fossil fuels and the CO_2 that breaks off limestone in all that heat to leave you with lime. And then there's steelmaking, mining, and all the other industries that enable modern life, all of which spew CO_2 to one degree or another. Even a solar panel is an expression of the fossil fuel economy, though over time it makes back the CO_2 involved in its manufacture.

Eisenberger seems to pick these fights, even among friends, perhaps to call more attention to him and his product (and preferred solution). As he notes, very few tax dollars are spent developing air capture technology while billions are spent on CCS projects at power plants, including Mountaineer in West Virginia, FutureGen in Illinois, and Kemper County in Mississippi in the United States alone. "Why is the current system operating in a way, spending all this money and allocating all this creative talent for technologies that do not solve the problem and, in fact, will extend the lifetime of the things causing this problem?" Eisenberger asks, his frustration finding its purest expression in the rapid vibrations of his fringe of hair. "I don't understand it."

But Lackner does. "A lot more money comes into this because there's a big industry that says, 'Protect me.' We come out from nowhere, so it's just harder."

That's not good enough for Eisenberger. "CO_2 goes out and we take it back out with air capture. It's not geoengineering. There are no scary consequences. It prevents our air from getting polluted. It's very simple." He has calmed down and the hair has stopped vibrating, but his voice has grown raspy. "We've adopted this weird mind-set that the simple and straightforward solution gets zero dollars because it's not ready, and the solution that will not solve the problem and is also not ready gets billions of dollars to try to make fossil fuels clean."

The amount of CO_2 garbage to store is gargantuan. As Lackner points out, if the world decides to aim for 350 ppm at some point, nearly 1 trillion metric tons of CO_2 will need to be pulled from the sky and put somewhere else. There seems to be plenty of storage deep underground in saltwater aquifers and porous rock, but that remains to be demonstrated. At best, people put CO_2 underground to flush out more oil, which does not do much to reduce CO_2 in the atmosphere since it comes right back up with the oil or is replaced by the CO_2 created when that extra oil is burned.

Demonstrations have proved hard to come by: Berlin stores explosive natural gas in the porous rocks beneath the capital, yet Germans have rejected plans to put CO_2 down there instead. Projects at sea have been more popular, including Sleipner, a natural gas field under the North Sea off the coast of Norway that now stores CO_2 as well.

"The case for coal cleanup is a lot of crap," Eisenberger goes on, undeterred. And basically saying the CO_2 capture for coal-fired power plants is a bad idea, though it is the idea that almost every scenario put forward to combat climate change relies on, at least in part.

"Air capture is the capture of last resort," Lackner responds with a smile, earning a snort from Eisenberger, who apparently scoffs at that notion. "Well, it is! If you have a power plant, you probably want to get CO_2 there. I don't see it as a competitor. First worry about coal plants and get rid of the CO_2, because it's the obvious first step. And let's do air capture in parallel, exactly in parallel." Even among physicists, energy quickly becomes tribal, adherents of nuclear fission against solar proponents, air capture absolutists versus all of the above CO_2 grabbers.

The ultimate goal is not to put any more CO_2 into the atmosphere, or at least pull back an equal amount of CO_2 for any carbon emitted. But the math to achieve that goal is daunting: More than 100 million of Lackner's proposed artificial trees would be needed to counter the nearly 40 billion metric tons of CO_2 spewed every year. For comparison, the world currently builds about 80 million cars a year.

Or perhaps another climate shift can be engineered with duckweed, not unlike Smetacek's scheme. Around 49 million years ago, a duckweed bloom atop the Arctic Ocean sucked in more than 1,000 ppm worth of CO_2 in less than a million years to end the Eocene epoch, scientists surmise, and ushered in the current low-CO_2 world. This tiny fern can double in number in five days. Way back then, it made a bloom that resulted in a layer of Arctic fossils 20 meters thick. It is this same layer that today's oil companies hope to tap for crude.

All this intellectual competition just shows that there are options, if and when the world decides to do something serious about CO_2 piling up in the atmosphere and trapping more and more and more heat. The problem is: The world has shown few signs of such seriousness.

Boreholes would reveal whether Oman's Hajar mountains might be made a huge repository for greenhouse gas pollution, a viable alternative to just dumping it in the sky. Quick calculations done on a scrap of paper suggest the mountains could hold at least 10 percent of the CO_2 spewing from fossil fuel burning, maybe more. But no one can be sure until Kelemen, or someone like him, tries it.

The idea is to drill the holes and dump CO_2-rich water down, allowing it to react with the vast hidden reserves of peridotite underground, far more than can be exposed at the surface. The process could be self-cracking, the reaction of binding the CO_2 in the first place fracturing yet more rock, which can then bind carbon.

And if the boreholes are deep enough, the heat from Earth's interior can speed the reactions. "You have a heat source and you don't have to grind the rocks," Kelemen says.

There are other ways, most notably the kind of weathering championed by the cranky Dutchman Olaf Schuiling, the kind of weathering that has kept Earth from suffering the same fate as its sister planet Venus, smothered in a blanket of greenhouse gases and experiencing surface temperatures near 500°C.

Instead, Earth has the Dolomites in Italy and the Cliffs of Dover in England. Throughout geologic time, the rains and atmospheric CO_2 have interacted with silicate rocks like the ones in Oman, and the resulting carbonated water washes down to the sea. In the sea, first microscopic plants use the bicarbonate to make armor—shells—a trick also adopted by everything from corals to crabs. Uncountable numbers of shells become a layer on the seafloor and, over the eons of geologic time, turn to rocks like dolomite and limestone. As Schuiling puts it, "These are the safe and sustainable CO_2 storage rooms of nature."

Chalk is a weird rock, the residue of the hard parts of trillions of ancient invisible algae. So much algae that chalk itself is almost entirely made up of such shells with nothing else mixed in, an unusual purity among rocks but common in the chalk layer that underlies much of Europe. Europe was once the bottom of an ocean and there was nothing to drift down to the floor of that long-ago sea except all those shells from dead algae, not even sand or mud. It is hard to imagine that the remnants of humanity will leave such a clear and massive mark, even if we turn to fertilizing the oceans.

But we have sped up the rate at which CO_2 goes into the atmosphere, so Schuiling, like Kelemen, believes that we must speed up the rate at which rocks take the CO_2 back. Schuiling plans to take mined olivine, grind it up, and spread the resulting sand grains on beaches and in the shallow parts of the sea—even in the sandboxes of nursery schools, he adds—to suck up CO_2 faster. The only thing that stands in the way is money—all the energy required to mine the rocks, grind them up, and spread them around must be paid for by someone. That someone has not stepped forward, though Schuiling is currently competing for the Virgin Earth Challenge's $25 million prize for a technology that removes greenhouse gases from the atmosphere.

There is also the matter of soil. Agriculture, which now spans most arable parts of the globe, has taken thousands of metric tons of carbon out of the soil to feed humanity's growing population. This mauling of global soils is reversible, and it would be beneficial both for agriculture and the air to reverse it. At present, agricultural soils release 20 metric tons of carbon per hectare per year. That carbon could be put back in the form of charcoal or biochar at the rate of nearly a billion metric tons per year for several decades. In other words, carbonizing the skin of the land could take a gigaton of carbon out of the air each year.

The trouble is that no one—neither farmers in Iowa nor emirs in Oman—wants to replace the atmosphere as the world's dump. "They're smart people. They read the newspaper. They don't want to be tarred with the same brush as people who are taking e-waste, for example,"

Kelemen explains with a shrug. "They don't want to be seen as taking out the garbage for the rest of the world."

Not unless there's a lot of money in it.

On an early spring day touched with the promise of warmth plus the threat of rain, I headed up to Columbia University's Mudd Building and Lackner's physics lab. His hearty chuckle belies his formal German diction and physicist's habit of hand waving with numbers. Girding myself for potentially indecipherable jokes, I'm here to see Lackner's potentially world-saving technology: a plastic that can capture CO_2 directly from the air.

The resin sits on the lab bench outside a clear tiny greenhouse that holds bamboo, basil, cucumbers, and a heart-leaf philodendron, all of which glow an eerie washed-out red under violet light. The plants' leaves rustle in the breeze from a Dyson bladeless fan. Next to the big tank, a computer monitor charts CO_2 levels, and a tube on one side separates the environment within from the lab outside. With the light on, the plants busily suck in CO_2 to make leaves, roots, and vegetables. "The cucumber got fat on the CO_2," Lackner notes and chuckles, perhaps relishing the thought of adding it to some future salad.

But when Lackner inserts a folded checker of pale beige polypropylene plastic embedded with 25-micrometer particles of the resin, almost immediately CO_2 levels inside the greenhouse begin a steady march downward as the resin binds the greenhouse gas to form bicarbonate, a kind of salt that serves the human body as a digestive aid and keeps blood from turning to acid. This salt, more familiar as baking soda when there's a sodium atom involved, holds the CO_2. The resin sucks in CO_2 even more powerfully than the plants do, depending on how dry it gets. The recipe for reversal, to get the CO_2 back out: Just add water.

This is no joke. The resin scarred a polycarbonate plastic bottle used to store it. "They broke the plastic," Lackner says, showing me the streaked, cloudy, hard plastic bottle. The resin pulled CO_2 out of the plastic in its vigorous quest for chemical equilibrium.

Lackner calculates that more than 700 kilograms of CO_2 pass through an opening the size of the door to this lab over a twenty-four-hour period when the wind is up, courtesy of another Dyson or just a windy building turing CO_2 from the air. He suggested CO_2 capture to her as a science project back in middle school in the 1990s. She solved the problem the old-fashioned way, using sodium hydroxide—also known as lye—and a fish-tank pump to pull CO_2 out of the air and lock it in sodium carbonate.

A bottle of lye—a primary ingredient of soap and the pulp used to make paper, among other products—must be kept tightly sealed. That's because the strong base will rapidly suck CO_2 out of the atmosphere if exposed to the air and undergo a chemical transformation into sodium carbonate and, ultimately, baking soda. The CO_2 can be reextracted by heating the baking soda above 900°C in a kiln, regenerating its ability to capture yet more CO_2.

With some improvements, that's basically the principle used in technology that keeps submariners and astronauts alive. The problem is that sodium hydroxide and other solutions are often dangerously noxious and do not want to give back the CO_2 so that the original chemical can be used again, and again, and again. Hence the need for 900+° temperatures. To break that bond requires too much energy, which is why an American Physical Society report suggested it would cost at least $600 per metric ton to capture CO_2 from the air. Nevertheless, Harvard's David Keith and others are working to improve the process.

Back in 2001, when Wallace "Wally" Broecker, the geochemist who named global warming, lured Lackner to Columbia to work on the idea, none of that was known. Lackner brought along his partner and better engineering half, Allen Wright, a refugee from Biosphere 2 in the Arizona desert, who still has a twinkle in his blue eyes when it comes to inventing CO_2-removal machines.

But it was clear even then that the solutions on offer probably could not do the job cheaply enough. So Lackner and Wright searched for materials that might do the same thing, stumbling upon the resin—Dow Chemical's Marathon A, typically used to purify water—through "dumb luck," in Lackner's words, the same method of error and fortuitous trial that Thomas Edison used to stumble upon tungsten as the right element for lightbulb filaments. With the resin in hand, they launched Global Research Technologies in Tucson. On the roof of the low-rise building that housed the nascent company, a pinwheel impregnated with the resin spun in the desert breeze, enduring the blazing Arizona sun, blistering heat, wind, dust, and even torrential rains. It survived, the first step on a

top. That's how much a sheet of this material might pull from the ai
it could be refashioned into a brushlike carpet, exposing even more o
resin and taking out even more CO_2.

Of course, 700 kilograms of CO_2 equals what thirteen people bre
out over the course of twenty-four hours. In other words, this resin w
need to become the most popular product in the world to have ev
small impact on pulling this trace greenhouse gas out of the atmosph
Lackner estimates that 10 million artificial "trees" would be require
drop atmospheric concentrations by 0.5 ppm per year—putting the gl
on track to get back to the 280 ppm or so extant at the dawn of the Inc
trial Revolution in a few centuries or so. Each machine would req
roughly 1.1 megajoules of electricity for pumping and compressing
kilogram of CO_2 captured. That's not to mention all the water requi
to wet the filters (and evaporate) in order to get the CO_2 out again
the resin can be reused to capture yet more CO_2. The compressed a
captured CO_2 could perhaps be used for industrial purposes, like sco
ing oil out of the ground to earn a little cash, or buried deep beneath
surface of the planet to keep it safely locked away. In other words, a v
industrial infrastructure of air-capture machines, pipelines, and co
pressors would be required to remedy the effects of our vast industr
infrastructure for fossil fuels.

And that's if the resin works as well as advertised. Figuring that out
the focus of the other experiment in Lackner's lab. Hidden inside a St
rofoam cooler—with a dark blue Columbia necktie as de facto latch—t
resin is exposed to water and CO_2 and is precisely weighed, and the ter
perature is kept constant. The idea is to keep the CO_2 steady at 400 pp
with no temperature variation and then change the other conditions
determine how well the resin really works.

One CO_2 molecule coming in seems to push out at least six wat
molecules, drying the resin further and cooling the area. Wet the resi
and it heats up. That suggests this is a technology that could work gre
in Tempe, the Atlas Mountains in Morocco, and the depths of the Austr
lian Outback, but perhaps not in Singapore or other humid climes.

Lackner started at Los Alamos National Laboratory with explosive
and fusion, nuclear weapons work, while dabbling on the side wit
thought-experiments on self-replicating machines. It was his daughte
Claire, now an astrophysicist, who inspired his present interest in cap

decades-long journey of development. But the start-up failed to prosper, even after relocating to San Francisco. There was little demand for the technology. Still, they have been able to test different versions of the resin made by Dow, Diaion, Rome and Haas, and other chemical manufacturers. "Everybody makes it," Lackner says. "All work."

These days, Lackner and Wright have returned to the desert to try to build a machine based on the resin at Arizona State University's new Center for Negative Carbon Emissions, a big step up from a pinwheel showing that it can pull CO_2 out of the air in the breeze. Even Broecker, the New York professor, had to admit that an artificial tree, even a small one, could never be built in Manhattan. So Wright has built a small prototype in the new Tempe lab, and taken it to the roof of the Interdisciplinary Science and Technology Building 4 for testing in the wind. If all continues to go well, he will build a full-scale artificial tree next to the Salt River Project's water flowing slowly through the desert. That might be in as little as three to five years.

There is plenty of magic technology in the world. An airplane can fly on sunshine across the United States or around the world. A small monolith that can make phone calls puts more computing power in the average pocket than all that was available in the 1969 mission that put men on the moon. The genetic instruction set for any kind of the exuberant profusion of life on Earth can be read by a machine the size of a microwave oven. But even Lackner admits there has been little progress on what may become the most important technology—a CO_2-based thermostat for the planet.

If carbon capture has not prospered, storage is even worse off. There are at least 650 billion metric tons of CO_2 looking for a home other than the atmosphere, though the U.S. government estimates that North America alone could hold 3 trillion metric tons of CO_2. And let's not ignore the problem of leaks.

No one wants to live near or over a garbage dump for the world's greenhouse gases. Horror stories seep from mind to mind. The volcanic Lake Nyos in Cameroon belched a CO_2 plume that asphyxiated people and livestock in 1986. At In Salah in Algeria, the ground swelled a full centimeter as CO_2 was injected. Fractures deep below were detected, completely unexpected and, perhaps more important, unpredicted by the best computer models and minds in oil and gas geology—and injection ceased.

There has never been a human project that didn't have some failure, whether launching astronauts into space or running nuclear reactors without meltdowns. "Never say zero release" is the caution from those with a background in the controversies over nuclear waste. Instead, note the maximum release and what dangers, if any, that release would pose.

Take the cautionary tale of the Utsira Formation off the coast of Norway in the North Sea, as much as 300 meters thick and sprawling over an area 450 kilometers long and 90 kilometers wide more than 600 meters beneath the seafloor. There are plans to store CO_2 there, but for the moment, the sandstone formation is where they dump the toxic water that comes back up with the oil and gas produced nearby. Statoil dumped and dumped, and before they knew it, a crater on the seafloor burst into existence, roughly the size of an apartment building. The contaminated water leaked into the ocean.

"This demonstrates why you don't want to store CO_2 under Milwaukee," Kelemen says. "But if that happened at Sleipner, you'd get a few million tons of CO_2 back, and unless there was a fishing vessel directly above, there would be no human impact. From a pragmatic point of view, even catastrophic events are not that serious." The worst-case scenario is that CO_2 that would have been in the atmosphere anyway ends up in the air.

Either there will be a breakthrough to allow fossil fuels to be burned without adding CO_2 to the atmosphere (and, it is hoped, remedy some of the other challenges, like soot and mountaintop removal), or the world will be stuck in endless promises of action like the Paris Agreement to combat climate change. "Evolution had 3 billion years to work on it, and it didn't get better," Lackner adds, with a wry smile, speaking of the version of air capture known as photosynthesis. "So it must be a hard problem."

April 2014 is the first month since before months had names that boasted levels of CO_2 over 400 parts per million in the atmosphere. Such CO_2 concentrations have likely not been seen since at least the Pliocene epoch, roughly 2 million years ago, and maybe even the Oligocene, 23 million years ago. That means we are breathing air never tasted by any of our ancestors.

Homo sapiens has subsisted for at least 200,000 years on a planet that has oscillated between 170 and 280 ppm, according to records preserved in air bubbles trapped in ice. Now our species has burned enough fossil fuels and cut down enough trees to push CO_2 to 400 ppm—and soon beyond. Concentrations are rising by more than 2 ppm per year now. Raising atmospheric concentrations of CO_2 to 0.04 percent may not seem like much, but it has been enough to raise the world's annual average temperature by 0.8°C so far. More warming is in store, due to the lag between CO_2 emissions and the extra heat each molecule will trap over time, in effect, an ever-thickening blanket wrapped around the planet.

Nearly 40 billion metric tons get spewed into the atmosphere each year at present, an amount that seems to grow inexorably. Yet the need to stop burning fossil fuels has been obvious since at least the 1950s. Earlier scientists, like the Swedish chemist Svante Arrhenius, who calculated by hand (correctly) how much fossil fuel burning could raise temperatures and, influenced perhaps by being located in coldish Northern Europe, thought that a little global warming might be a good thing. By 1959, the physicist Gilbert Plass could write in the pages of *Scientific American* about the growing peril posed to the planet's climate by carbon dioxide. A few years later, in 1965, a report on restoring the quality of the environment landed on U.S. president Lyndon Johnson's desk and suggested that spreading reflective particles across the surface of the ocean might counterbalance the expected warming, the first serious suggestion of such large-scale interventions in planetary systems. By 1977, one of the authors of that report, the geochemist Wally Broecker, could say, "I don't think there's anything we could do to stop it from going above 450." He added later, "We're not going to know empirically until at least 2010, and by then we will have made a decision, I think, about how much coal we're going to burn."

We are past 2010 and have indeed made that decision—as much of it as can be mined from the ground. We are well on our way to 450 ppm, and perhaps beyond, so the need to get off fossil fuels only becomes more apparent with each passing year.

We aren't done yet. Greater concentrations will be achieved, due to all the existing coal-fired power plants, more than a billion cars powered by internal combustion on the roads today, and yet more clearing of forests. That's despite an avowed goal to stop at 450 ppm, the number broadly (if infirmly) linked to an average temperature rise of no more than 2°C over the next century. More likely, by century's end, enough CO_2 will have been spewed from burning long-buried stores of fossilized sunshine to increase concentrations to 550 ppm or more, enough to raise average annual temperatures by as much as 6°C in the same span. That may be more climate change than human civilization can handle, along with many of the other animals and plants living on Earth, already stressed by other human encroachments. Such a shift will certainly be too hot for the great ice sheets, and the seas will inexorably rise, drowning coasts. The planet will be fine, though; scientists have surmised from long-term records in rock that Earth has seen levels beyond 1,000 ppm in the past.

In fact, by human standards, February 2015 marked the dawn of a new climate, the 360th of a 360-month span all registering above-average temperatures. Since climate scientists measure climate over increments of thirty years, that means global warming is the new normal. We live in a changed climate, another sign of the Anthropocene.

For Lackner, it is too easy to imagine relatively small climate swings that would be too much for human civilization to handle. We have a bad track record of dealing with even minor climate change—smaller shifts have ended Chinese dynasties and Mesopotamian empires—though we have better technology now. Still, he foresees a problem in what he calls wet bulb temperature, or what enables sweat to cool animals like us. This is the lowest temperature possible given cooling by evaporation, and pretty much all over the world it never exceeds 31°C. If it goes much higher than that, sweating will no longer shed heat, and humans (and other sweating mammals) will be in deep trouble, as evidenced by the deep mines where this already happens and so miners must wear jackets filled with ice. If the wet bulb temperature goes above 35°C, humans will be trapped inside on our planetary home, reliant for life on air-conditioning and doomed if the power goes out. That would take only 7°C warming of average temperatures, or what might happen by the end of this century if present pollution levels swell too much.

Adaptation is the name for the climate change coping strategy that supposes it's too late to do much about rising atmospheric CO_2 concentrations so we might as well grin and bear it, do what is necessary to adapt. After all, even if air capture works, it won't be a quick fix. It takes a long time for changes in CO_2 concentrations to translate into changes in climate, as the twentieth century and the early years of the twenty-first have proved. The lack of action at the UN Climate Change Conference in Copenhagen in December 2009 was enough to convince me that our adaptable species will be pressing our luck with adaptation again. I am not alone in that opinion, even with the hopes created by national commitments to action under the terms of the Paris Agreement of 2015.

"People will overshoot or have overshot truly sustainable atmospheric CO_2 concentration, and the natural drawdown is really slow," Kelemen notes. "Probably, people will decide to resort to negative emissions."

Think of it as a sewer system for the atmosphere. Prior to the nineteenth century, people in Europe dumped the contents of chamber pots in the street, tossing shit and piss out the window and onto unlucky pedestrians. The waste gathered in enormous open-air cesspools, and disease, especially cholera, was rampant. In 1821, London, the largest city in the world, had no enclosed sewers.

The lack of rain that year meant that all that human waste piling up in the streets, along with horse dung and other foul leavings, did not get washed away, stench and disease accumulating. It got so bad that Parliament considered leaving London. Instead, it passed a law, and within six years, underground sewers had been put in, digging up all the city's major streets. It cost about 2 percent of the United Kingdom's gross domestic product, by modern calculations. "Two percent of GDP buys a lot of carbon capture," Kelemen notes. "Perhaps all the carbon capture you would ever need to do."

One problem is that CO_2's unpleasant side effects have only begun to be dimly discerned, unlike the immediate negative impact of raw sewage. Another is the sheer scale of the problem: at least 10 billion metric tons of CO_2 captured and disposed of each year. That's an infrastructure and an industry bigger than iron mining for steel (1 billion metric tons or so a year), oil (4 billion metric tons), or even all the coal, oil, and natural gas in the world. By 2100, we might need to bury as much as 40 billion metric tons of CO_2 per year. It is far more than a mountain, even a mountain range in Oman.

But it is nothing compared to photosynthesis. Plants, algae, and other green life annually suck up ten times that amount of CO_2. And plants could theoretically do more if people could just help them use more of the energy in sunlight to bind carbon rather than fuel other processes (so much energy wasted on life).

There are other ways of messing with the atmosphere, which is a weird place. What is now called the Bodélé Depression (and the "dustiest place on earth") used to be Lake Chad. Persistent droughts in the Sahara Desert and increased demand for fresh water caused the lake to disappear. Now, frequent dust storms, full of plant nutrients, fertilize the Amazon rain forest a continent and an ocean away. Microbes and pollen in the air help it rain or snow, and there is a whole culture of microorganisms wafting around on even the highest breezes. The sulfate particles put out by a big volcanic eruption can cool the entire globe, as in the case of Tambora in Indonesia in 1815 and Pinatubo in the Philippines in 1992.

That has given some scientists ideas. By mimicking volcanoes and spraying a thin layer of sulfuric acid high in the sky, the earth could be cooled. This would work like a sunshade, blocking sunlight, but at the cost, perhaps, of a diminished ozone layer to protect life from the sun's ultraviolet rays and thus more deaths from air pollution, including asthma, heart disease, and lung cancer. Nevertheless, such solar radiation management represents a strong temptation to some.

There's also the fact that it could be done for roughly the price of a Hollywood blockbuster movie, just a billion or so dollars and a fleet of retrofitted airplanes, plus a ready source of sulfur. The refiners of Canadian tar sands have already been building giant yellow pyramids of the element in the boreal forest of Alberta.

Geoengineering of this kind is cheap and easy, which leads to the dreams of rogue geoengineers like Russ George, or even just a random billionaire with the means to change the skies. He or she might need to deflect only 10 percent of the incoming sunlight with a sulfuric acid sunshade to put Earth into the snowball scenario depicted in the movie *Snowpiercer*.

On the flip side, we could continue doing what we're already doing, but to excess, to boil away all the oceans and become another Venus at roughly 30,000 ppm of CO_2, and much lower really once more powerful, synthetic greenhouse gases are enlisted, like the hydrochlorofluorocarbons we use in air conditioners, or the sulfur hexafluoride that keeps the electric grid humming. Perhaps some future group of humans will build a crude spaceship, load it up with such chemicals, and crash it into Mars to kick off massive global warming there and prepare the way for some future habitation.

Or we may end by turning Earth into Edgar Rice Burroughs's Barsoom: a planet kept alive by an atmosphere factory where life takes place in domed, dying cities whose inhabitants have access to technology that preserves lives for at least a millennium.

The problem with the illusion of control is that it can lead only to more efforts at control, not unlike a drug addiction. Once we start to mess with the skies, we won't ever be able to stop. That's been the analysis of the best minds in the United States since the 1950s, and I'd add that that kind of technological change to the skies would have a profound impact on political and social relationships down here on the ground.

We don't blame witches for weather making and then burn them at the stake anymore. Then again, that day may yet come for would-be geo-engineers, or maybe coal barons.

There is another way: to stop befouling the sky in the first place. No matter how much haze gets pumped into the stratosphere or marine clouds get brightened, CO_2 has to be drawn down. There are enough fossil fuels around to raise CO_2 concentrations beyond 5,000 ppm, but it takes only 1,200 ppm or so to start dissolving coral reefs in the ocean where they stand. It's time to take more care.

A mild late summer day, the air redolent with the smell of this mixed forest of oak and scrub heating up in the morning sun, along with cut grass and the mineral tang of mud and clay. Just off the Palisades Parkway, perched atop the sheer cliffs that give the highway its name and just across the New York–New Jersey border, a relatively small drilling rig hangs off the back of a diesel truck jacked up on a couple of wooden two-by-fours. The rig itself just clears the surrounding trees, poking up into the sky like an advertisement for oil, and the diesel engine of the drill looks about as old as the Age of Oil itself. A couple of big metal tubs—looking ready to bathe some cowboys—hold muddy water for drilling, and pale gray core discards litter the forest edge, along with rusting lengths of drill pipe.

The idea is to drill into the sandstone and diabase formed 200 million years ago as the Atlantic Ocean began to open. The Palisades that line the Hudson River across from Manhattan are not ramparts of some stone giant's fort but rather the remnants of ancient lava flows. The drillers pull out banded cores, the scientists score it with a red marker along its length, occasionally stopping to denote in black marks the number of feet down, 897 to 901 covering thousands of years. The cylindrical rock turns jagged where one core ends, a break, and then another core begins. The two chunks fit together like jigsaw puzzle pieces, albeit ones that may change color from mottled black to cement gray in the span of a single foot, or even a single inch.

The goal is 2,000 feet, each inch another bit of history pierced, and fresh cores rest under a tent. "Two thousand feet or basement, which-ever comes first," jokes Paul Olsen, a fit, gray-haired grandfather and geologist from Columbia University, who wears glasses and dresses all in black but for his white hard hat. The basement in this case is the point at which the hardened parts of the mantle create a floor for all the sedimentary rocks above. The mottled red, black, and gray cores are stored in three-foot chunks, and one core might cover the Triassic—the first age of dinosaurs—to the Jurassic, 200 million years ago, more or less. The goal is to find out just how porous solid rock can be, in the hopes of squeezing millions of metric tons of CO_2 molecules into the deep earth.

Diabase is a tough rock, the kind of magma that never made it to the surface, with few cracks, which is why they make roads out of it, like the crushed gravel that the drill rig truck sits on. This parking lot is a CO_2 sink, though the gray rock does not look very special, just dense. Diabase has few pores to put the CO_2 into, and it is nearly impermeable. But if cracks can be made, perhaps using the same fracking used to free natural gas, diabase can react with CO_2 to form a kind of limestone. So as more and more CO_2 is pumped underground and more and more time passes, more and more limestone is formed. Or so goes the theory.

Sandstone, on the other hand, has a lot of nice holes in it for storing CO_2, pushing out the water that often sits there now. Like the water, that CO_2 will just sit there, kept in place by a mudstone cap above the sand-stone. But, like a cork always trying to pop to the surface when pushed under water, that CO_2 will be waiting for a crack or an earthquake or some other event to set it free once more.

Long ago, the Palisades were an immense rift valley, like the one in East Africa today. The rift extended from today's Greenland down into the Gulf of Mexico and reflected the split between continental crust that would become North America, South America, and Africa. What is now temperate New Jersey sat on the equator way back then, and the rocks show a cycle between lake beds, dryland, and rivers, judging by the red mudstones that reflect riverbeds and the black mudstones for lake bottoms. "What we are seeing is purplish," Olsen says. And then, roughly 195 million years ago, this great rifting stopped, geologically overnight. "We don't know exactly why," he admits.

The top of the core, 700 or so feet down, is Kelemen's beloved olivine. Then there's diabase. They plan to test injecting CO_2 in water itself, basically shooting Perrier back underground. Similar tests are under way in Iceland, the western United States, and even Australia.

This magma that protruded up is 300 meters thick and 500 kilometers long, at least. It's not known if it's contiguous with similar magmas found up and down the East Coast, but it is similar to lava flows in the Deccan Plateau of India and the Siberian Traps (both vast regions of volcanic rock), just more eroded. And like those magmas, this rifting may have caused extinction, either from escaping CO_2 or the simple burning of hot lava itself.

There are many unknowns about drilling into diabase because the oil companies usually avoid it—intruded magma is believed to cook away any valuable hydrocarbons that might be lying around in the formation. Nevertheless, this shovel-ready project got some of the American Recovery and Reinvestment Act money that the U.S. Department of Energy provided in hopes of charting a future for coal. Public resistance is likely to be fierce. "If you're not happy with windmills, you're *really* not going to be happy with sequestration," says the oceanographer Dennis Hayes of Columbia, another member of the drill team and the one expert in ocean crust.

This whole operation could be shifted offshore, where there is no one to trouble, like Sleipner. What is gained in safety, however, is more than made up for in the expense of drilling under the sea and building pipelines.

It remains to be seen what kind of chemistry these magmatic rocks are capable of. If the rock came direct from the mantle, it might be able to hold lots of CO_2, and if not, not, a prospect that gets Olsen up on his heels and waving his arms—an argument among geologists over the story the rocks tell.

These cores will ultimately find a home in the repository of the Lamont-Doherty Earth Observatory's geosciences building, a squat, dark-brick hulk more than thirty years old and showing its age. Half the cores in the collection look as if they have been bored by some burrowing sea creature—really just generations of scientists taking samples—and half remain as pristine as when they were drawn out of the ground, a record of sorts, waiting for future techniques to reveal yet more secrets.

You can't just bury the problems of coal; there's also cleaning up the mess from all the burning of fossil fuel to date. For that we'll need Lackner's artificial trees, or some other magic device. There's also the problem of who controls the thermostat, like a fight over the ambient temperature of the LDEO geosciences building but scaled up to the size of the planet. Sweden might prefer a bit more CO_2 to warm things up, as Arrhenius surmised long ago. On the other hand, Australians might think it's quite hot enough, thank you, given that recent heat waves have forced them to add new colors denoting ever higher high temperatures to their weather maps.

And then there is all that remains unknown. "How can you correlate atmospheric circulation in Arizona with precipitations in Somaliland? Or wind pressure in Bhutan with vegetation change in Brazil? At what confidence level? Here the probability is small if not nil," argues the geographer Cush Luwesi of Kenyatta University. "If we cannot do that, whatever climate intervention that will be put in place in a region will improve one aspect of the climate in that specific region and worsen other variables therein and elsewhere in the globe, depending on the spectra of its impacts." Who will decide, and how?

So this team from Columbia drills into the Palisades, looking for a proper place to store the CO_2 billowing into the sky after all the burning: 300 metric tons of coal each second, 3 billion barrels of oil each month, and 3 trillion cubic meters of gas a year. And that's with more than a billion people still not using much in the way of fossil fuels.

The world needs plenty of new ideas—a cheap, clean cookstove to save the poor from breathing smoky air that kills, a better way to contain the sun's fusion here on Earth, a method to reuse the king's ransom of precious metals in any given mobile phone—but CO_2 capture and storage is not one of them. There are enough ways to catch the stuff, from Kelemen's mountains to Park's specialty molecules, and ways to store it either in useful products like fuels or deep underground. The ideas are there, but the proof that the ideas are good is not. Because proof takes money.

Oman is no Saudi Arabia, but it does have oil and gas and plenty of money, either for drilling or for hosting a solar plane that took off for a worldwide voyage to prove that it could fly around the world with no fuel. But it does not necessarily want to be the world's garbage can for CO_2.

Still, the rock of Oman holds promise. When heated to roughly 185°C, the rate at which olivine will absorb CO_2 becomes a million times better. That's 1 billion metric tons of CO_2 per cubic kilometer of rock per year. The potential is there to get rid of a lot of excess CO_2 in these unprepossessing rocks weathering away in the desert rains year after year, spending a geologic eye blink—roughly 26,000 years by Kelemen's calculations—at the surface.

Drilling can get to rock that's already at 185°C. Those rocks are a mere 4 kilometers down, not even half as deep as the ill-fated Macondo well drilled by the Deepwater Horizon in the Gulf of Mexico. And this hole could start on land. "That's starting to cost you, though," Kelemen admits, with another wry smile. "What you really don't want to do is pay to heat all the water you inject."

Still, it's something that, say, the geothermal industry has plenty of experience with, or, as Kelemen puts it, "A monkey with a lever could regulate this. It wouldn't really be that complicated . . ."

"Good," I interject, "because it will be apes with levers regulating that."

The problem is scale or, as Kelemen notes: "You end up needing a lot of holes." He adds, "Any scale of capturing and storing CO_2 looks pretty appalling to the layperson," something on the order of as many oil and gas wells as have been drilled to date in the Unites States—that's millions of wells.

First, it will have to be proved in the rock. So Kelemen hopes to drill a hole in Oman—not to dump CO_2, just to listen. By lowering seismometers into the subsurface, Kelemen might be able to detect the self-propagating cracking. "I'd like to hear that happening," he muses. "Nobody knows what's going on down there."

There is no Department of Negative Emissions, or even a global treaty to say that emitting CO_2 is bad—the Paris Agreement didn't go quite that far. Power plants, oil refineries, and other appurtenances of our fossil-fuel addiction last for decades and it's expensive to replace them or clean them up. We're still building more. And the fossil-fuel industry and some fossil-fuel-producing nations seem incapable of admitting there is a problem, even actively funding campaigns to obfuscate the issue.

We need some kind of carbon capture because the global civilization we have built continues to add more and more CO_2 to the atmosphere, with no signs of slowing, other than the occasional depression in the United States or China. Behind China, India is waiting for its turn to pollute, as are the 2 billion or so people who do not benefit from the burning of fossil fuels but bear all the consequences of climate change: drought, famine, infectious disease, dirty air. "We cannot support 10 billion people on this planet in reasonable comfort without lots of energy," Lackner notes. Civilization currently uses 14 terrawatts of energy in a year and may be headed to 30 terrawatts or more over the course of this century.

The world has wasted thirty years that could have been spent developing techniques and technologies to solve the CO_2 problem. The world remains hopelessly addicted to the cheap but dirty energy embodied in fossil fuels. "We're probably going to burn all of it," argues John Hansen from Denmark's Haldor. "It's stupid, but we're going to do it. What happens when we have burned it all?" The world will be in desperate need both of climate fixes and new ways to get energy.

All hope is not lost. Even Eisenberger is not as skeptical as some of his remarks seem to suggest. "If you look back on the history of humanity, how quickly can you change something as fundamental as we are trying to change here, we are actually moving very quickly. It's just we've gone down a lot of intellectual dead ends."

In the end, Kelemen's idea is the one people react to most favorably. Perhaps it's like the marketing of food—people are more comfortable with something that seems natural rather than something engineered. This is a natural system for taking CO_2 out of the sky, and we are going to need it, and probably all the others too. Plus, there is olivine and serpentine lying around outside nickel mines, already ground up and ready to go.

The best part of turning CO_2 to rock is that it eliminates the need to keep an eye on it. The same is true if the CO_2 can be transformed back into a fuel or other product, like the big blue-green foam C, O, and 2 left up on the Columbia dais and made from carbon dioxide turned back into plastics. That said, a coal-fired power plant emits 1000 metric tons of CO_2 every second. Nobody needs that much Tupperware, though it does imply considerable job security for this conclave of scientists.

The rock is there, waiting in the Hajar range. It probably cannot be moved, but perhaps someday machines to suck CO_2 out of the atmosphere could cover those ridgelines like a row of wind turbines and bring the CO_2 down to the rock even faster. We could engineer another Carboniferous period. Otherwise, we're headed into an Anthropocene thermal maximum.

Chapter 8

The Final Frontier

There is a sort of poverty of the spirit which stands in glaring contrast to our scientific and technological abundance. The richer we have become materially, the poorer we have become morally and spiritually. We have learned to fly the air like birds and swim the sea like fish, but we have not learned the simple art of living together as brothers.

—MARTIN LUTHER KING JR., NOBEL LECTURE, 1964

The goal was to meet the great man himself, if I'm honest. I planned to travel to the Bay Area and, given past coverage and future prospects, I thought a guided tour of Tesla's newly refitted car factory might be of interest, perhaps along with a test-drive. And who better to be that guide than Elon Musk?

I had just seen the inspiration for Robert Downey Jr.'s Tony Stark character in *Iron Man* onstage in New York City. He had been accepting an award for an automotive first: An all-electric car had been named best car of the year by the gearheads at *Motor Trend*. Not best electric car, not best alternative vehicle. Best car. Period.

In just ten years, Tesla had gone from being derided as the next Tucker or DeLorean to producing one of the best cars in the world. The judges cited its drag-race-ready pickup, spacious interior, and the fact it basically had a giant iPad in the center of its dashboard to control all your in-car needs, whether audio entertainment or GPS. And the Model S won despite a litany of glitches, including an inability to get into the car when the pop-out door latches fail to pop out. "We've made a lot of mistakes with the cars," Musk admitted at the event. The key is to quickly recognize the mistake and rapidly fix it in the next car off the line or through a software update sent invisibly through the Internet and Wi-Fi.

Fail often but fail fast—that's become something of a mantra in Musk's natural environment, Silicon Valley. And the key to Musk's success has been providing a must-have accessory in the form of a car to his fellow tech bros, newly enriched by Google, Facebook, and a seemingly endless litany of apps. The Tesla Model S in Palo Alto has become like the Toyota Camry everywhere else, simply the most popular car. Bonus: the satisfaction that comes from the fact that driving a Tesla can cut down on the greenhouse gas pollution warming the world. After all, nothing comes out of the tailpipe (those emissions displaced to some faraway smokestack, at least); indeed, there is no tailpipe. Musk's stated ambition is nothing less than saving the world.

To that end, Musk has made Tesla's patents open access, encouraged in that gesture by the fact that few of his competitors buy parts from him. He has urged "greater vigor" in the pursuit of electric cars from the General Motors, Toyotas, and Volkswagens of the world, behemoths that churn out thousands of vehicles a year, and attempted to craft an affordable electric car with the Model 3. On the other hand, when it comes to luxurious electric cars, Musk has had the field mostly to himself.

Yet somehow despite all these accomplishments, Musk's persona on that New York City stage was the opposite of confident, a bit mumbly and shifting from side to side as he talked to the crowd of a hundred or so glitterati, twisting this way and that as he strove to talk up Tesla. But I thought I could detect an undertone of arrogance in his robotic demeanor. Here's an example of a Musk joke. During the 2012 presidential campaign, Republican candidate Mitt Romney had derided Tesla as another example of an Obama administration clean energy "loser," like the failed advanced solar panel manufacturer Solyndra. "He was right about the object of that statement," Musk quipped, "but not the subject."

Who is Elon Musk? His name sounds like the overheated output of a sci-fi novel written by a computer, or perhaps a kind of high-end perfume. He might be an alien. Like me, he once liked to play Dungeons & Dragons—a fantasy role-playing game involving dice, percentages, and an overactive imagination—most often running the show as the Dungeon Master. The DM is the person who dreams up the entire adventure that the rest of the players experience. In a sense, Elon is now the DM of the world that the rich, at least, are living in, a world of electric cars, journeys into space, and ubiquitous solar power.

Elon Musk is not afraid to say he wants to bring mammoths back, whether it's wrong or not, and also has enough money to assay geoengineering if he ever decided the world needed an emergency fix for global warming. He even has the obligatory private jet or, even better, the rocket fleet. Perhaps that's why he has poked fun at himself on Twitter with a picture in which he is cradling a stuffed animal version of a white cat and holding a pinkie up to his lips like Mike Myers's Dr. Evil. "I want to be on the cover of *Rolling Stone*, that would be cool," he told CNN back in the late 1990s, after he made his first few hundred million. That ambition actually seems beneath him now, though he does keep framed copies of his appearances on the cover of *Aviation Week* at his SpaceX offices.

At the after-party as the crowd milled around the best car, I spotted what looked like Musk hiding on a couch in a dark, back corner. I assumed it was a clone, or at least a brother, since Elon was no doubt in high demand at the event. I also would not put it past Musk to clone himself, despite his musings on the potential dangers of genetic engineering, if only to be in two places at once—a seeming necessity when running two companies outright and advising a third. When I approached and tried to chat, however, the clone demurred. It was only later that I realized it was Elon Musk himself, doing his work by himself in a far corner and merely observing the party, chatting mostly with his mother.

So I had come close to meeting the man before. Even better, his PR folks seemed to think a Musk-guided tour of Tesla's factory would not be a problem.

The problem it turns out was going to be Musk himself.

Much of Musk's personal fortune—and time—is invested in infrastructure: the SpaceX hangar in California, a new lithium-ion battery Gigafactory in the Nevada desert, and the largest automobile factory west of the Mississippi River in Fremont, California, in the southeast corner of San Francisco Bay. The sprawling building, big enough to hold all of the Vatican or most of the Pentagon, started as a General Motors facility to serve the burgeoning car culture of California and the western United States in the 1960s. In the 1980s, as the American car colossus declined, it became a joint venture between GM and Japanese upstart Toyota with the bland yet ominous name New United Motor Manufacturing, Inc., or NUMMI (pronounced "KNEW me"). The past tense is appropriate, as the joint venture failed—GM having been unable to digest Toyota's manufacturing secrets and the Japanese company finding cheaper and better labor elsewhere in the United States. Like a bad joke, the factory closed on April 1, 2010, felled by the Great Recession.

Musk saw this as the fire sale of the first decade of the twenty-first century, especially after the U.S. government had rescued his fledgling Tesla car company from a similar fate with a loan of $465 million in 2008. By May 2010, Musk had bought the facility for $42 million, practically highway robbery for a factory that covers 5.1 million square meters. Rechristened the Tesla Factory, it became a point of pilgrimage for evangelical owners. Much as Mercedes owners flock to Stuttgart to pick up new vehicles, those with the means to purchase a Tesla often prefer to take delivery fresh from the factory. As one Tesla employee told me during a tour of the factory on a hot day in Fremont, "Our customers are our best salespeople."

Despite that popularity, there is a dark side to the Tesla Factory, literally, as I found on my visit on a sunny day in early summer. More than half the facility remains dark as Tesla struggles to pump out tens of thousands of vehicles a year, compared to the 300,000 vehicles the NUMMI facility once produced.

The part in use is bright and clean, the floors lacquered white on Musk's orders. It's "for pride," he explains to employees, and an aesthetic carried over to other Musk facilities, though it makes cleaning difficult.

Giant KUKA robots swivel and pivot, slowly manufacturing cars as if constructing an extremely repetitive LEGO set, supervised by a skeleton crew of actual humans. Each robot bears the name of a superhero— Wolverine, Cyclops, Iceman—and all can swap their own "heads" for different jobs, like riveting or welding. A facility that once employed tens of thousands now hosts 4,000 survivors, working on one of just two shifts.

Right next to the production line sits Musk's desk, a bicycle propped against it to facilitate quicker travel around the sprawling factory. On the day I finally visit, the desk sits empty, Musk called away to SpaceX. I will not meet the man himself today.

Tuesdays and Wednesdays are Musk's typical Fremont days, per his insane schedule mapped out on a whiteboard and no doubt also hosted in the cloud somewhere. His Tesla employees sit out in a bullpen-style arrangement of desks, mere meters away from the relative quiet of the robots at work. Some have bikes like the master himself, and others zip around on little scooters. Musk has an unimpeded view of the robots at work, or he can climb a little viewing platform and gaze down on the bobbing tips of the robots over the cars, like giraffes lowering their lips down to a watering hole.

In the back of the factory, five giant hydraulic presses—the largest in North America and bought used again at bargain prices—stamp aluminum into shape for the Model S, all in a hissing row. The presses extend four stories up, but peering into their innards from a safe distance I can see they extend another few stories underground as well, slowly crushing out complicated shapes like hoods and side doors. The slow crushing means no heating of the aluminum and thus no warping, preserving strength. Laser cutters precisely machine metal parts, as Musk's demanding nature (and Tesla's small volume) require that most parts be made right here.

The slowly assembling cars glide silently on smart carts that follow magnetic strips around the factory from station to station. As I pause to look more closely, a robot spins, picks up a roof, hoists it into the air, rotates again, and delicately, ever so delicately, slots the roof into place with precision, almost as if it had fingers.

Each car is basically custom-built for each customer, hence the devotion, pilgrimages, and cost. Plus, safety. The lack of an internal combustion engine means that each Model S has basically three times as much material to crumple before car parts reach human parts in an accident. Liquid cooling of the batteries makes overheating—thermal runaway, in the lingo of engineers—unlikely, despite using the same standard Panasonic cell batteries found in more combustible electronic gadgets. All those batteries on the floor of the car make for a very low center of gravity, just 44 centimeters off the ground. The car is hard to flip, which is a good thing because it's fun to drive fast with its instant acceleration and light weight.

At the end of the line, the cars roll out onto a bamboo floor where humans check the robots' work. Then it's into the water-test chamber, which simulates a hurricane to make sure there are no unexpected leaks, a major no-no in a car that runs on electricity.

Driving the Model S is a trip into the future, albeit a very, very expensive one. Hence this particular future's uneven distribution. No detail is overlooked, including, as legend has it, the uncanny ears of legendary music producer Rick Rubin personally tuning the audio system.

The car glides silently through traffic. Acceleration is instant, as is deceleration, and braking recharges the batteries. The hills east of Fremont that challenged my little four-cylinder engine rental car are as nothing to the Model S. It's dangerous how easy it is to get going fast, even uphill, slipping into speeding without thought or effort. It's like driving a computer, and it is easy to see how such driving could be automated. Musk thinks that future is just a few years away, as is the danger that the artificial intelligence being baked into the computers capable of driving might prove an existential threat.

That would be a shame, however, as the car is so much fun for a human to drive.

The car whips around turns, a firm grip on the road thanks to that floor of heavy batteries. The sun bakes the California hillsides brown, and the redheaded condors so laboriously restored to the wild swirl high in the air on thermals. They keep a beady eye on the lazy cattle far below. It feels good to zoom past stations advertising gasoline at $4 per gallon. To go 95 kilometers on twisting hilly roads uses up 22.4 kilowatt-hours of the 85 kwh battery pack. You may not be able to guilt people into buying an electric car, but Musk's car proves you sure can seduce them. As I park in the lot of a local In-N-Out Burger for lunch, a knot of teens, boys and girls, comes over to find out if this is the Tesla they've been hearing so much about and snap pictures with their cell phones. A grasshopper stands out bright green on the deep red of this borrowed Model S, showing the attraction is not confined to just one of Earth's species.

At a charging station in a nearby mall, other Tesla drivers congregate to swap tales and jostle over the two cords for the five or so cars looking for some juice. "You can zip around, but it's almost better to just comfortably drive at the speed limit," offers one new Model S driver. He wonders if his car will be able to handle the elevation and cold at Lake Tahoe, and the rear-facing backseat has proved a little too hot for his kids. "A little tint or angled AC vents can take care of that," he suggests. Such chatting is known to the Internet as "Tesla time," and makes the minutes spent charging speed by, as does checking email on a smartphone. By my count, it takes about twenty minutes to gain back ten miles of range with this (slow) charger. That's nothing: Anthrome ecologist Erle Ellis snakes a long orange electric cord across his Baltimore lawn to plug in his Prius curbside overnight. On the other hand, "gas engines are a kind of Rube Goldberg contraption, it's shocking it even works," Musk has said, given the controlled series of explosions required, among other engineering miracles. "One day we'll look back on gas cars the same way we look back on steam locomotives. No one will actually drive one."

The real problem with the Model S might be the lead foot you discover you have acquired when you get back into one of those traditional cars—and the fact that the basic model costs $50,000 and doesn't even solve global warming unless the electricity is coming from the wind, sun, hot rocks, or fission. As one of Musk's employee's jokes to me on the tour, he owns a Roadster because he loves fast cars, works at Tesla, and "makes poor financial decisions." The Model 3 aims to ameliorate that problem.

I had wanted to talk to Musk about that, but no dice. And when I called back a bit later in the year, even proffering the opportunity to take part in a documentary all about electric vehicles, yet another PR person told me that Musk was not doing any more interviews that year.

That's too bad. I have a lot of questions Musk might be well-placed to answer, or at least have interesting speculations to offer. Between photovoltaic panel financier and installer SolarCity, electric carmaker Tesla, and SpaceX, Musk is inventing—and selling—a specific future. He has even invested in the kind of ocean carbon credits that Smetacek worked on, back in the days of Russ George's Planktos venture. Musk could well become the kind of Bond villain one might call "Greenfinger," and he has joked about treating himself to a volcano lair. In fact, Elon Musk has the resources to pull off another longtime dream of many an aging white man: dying on Mars.

July 16, 1945, was the first day in the history of the world with two dawns. The first one people made. It came at 05:29:21 in the darkness just before the dawn of a New Mexico morning in the White Sands desert best known for the trail called Jornada del Muerto, Journey of the Dead Man. The mountains and sprawling valleys of this region are as empty in form as in imagination and therefore the perfect spot to test a secret, world-destroying new power.

A real sunrise in the desert offers a sky smeared with pink clouds drifting in a lightening blue sky, accompanied by a chorus of birds. But not on July 16, when a flash of light and a deafening roar heralded the world's first mushroom cloud—and, potentially, the dawn of the Anthropocene. "Up 'n atom" may have been the slogan for kids at the dawn of the Atomic Age, but it would have served just as well for the world-famous scientists watching that predawn atomic test.

The physicist J. Robert Oppenheimer and his crew detonated an ugly little sphere sheathed in cloth and dangling from an Erector Set–type tower that looked not unlike an oil derrick. Scrawny geeks in T-shirts with pencils behind their ears tinkered with the test bomb's innards before it was hoisted, carefully, ever so gingerly, to its final position to rest overnight. There was the slight chance that exploding this bomb could kick off a chain reaction that would consume the entire atmosphere. Implausible by the physicists' calculations but not impossible.

Even film of the test can't quite capture it, as light blazes and whites out the view before slowly fading back to black in the silence of early film. In reality, the darkness was rent by light and heat and dusty air, kicked into the sky. That first, false dawn was closely tethered to the ground.

The Trinity site looks the same today as it does in the Department of Energy footage of the preparations for the test. Tumbleweeds skip and hop across this dry and dusty land, this invasive plant from the Russian steppes piling up against barbed-wire fences or imploding under the wheels of a passing truck. The air is so dry it sucks the moisture right out of you, and there is no noise other than the wind and the occasional bird scared into flight. A miles-long train crawls along the ridgeline of the valley of the Rio Grande, the river that brings life to this desert and explains the existence of Albuquerque and Truth or Consequences.

Dust storms scour the lowlands, but water flows in the bed of the big river. It is a stark green line in the land. Black and brown cattle graze endless fields of burnt yellow grass, spotted here and there by telescoping shrub palms. The hills and mountains that mark the horizon slump slowly in the rain, turning the Rio Grande brown with silt.

I suspect dust storms may outnumber the rains here in New Mexico. And what is in the dust picked up by the wind off this particular patch of ground in the high desert? Or the water in Albuquerque or Truth or Consequences? There is now something that wasn't there when earlier inhabitants carved the petroglyphs, wasn't there before that July morning in 1945. After that, it's in my teeth, my wife's teeth, my kids' teeth, and will be in my grandchildren's teeth if I am so lucky as to have them. They call it the bomb curve, a legacy of the many nuclear explosions that followed that first test through the long years of the Cold War and right up until the end of the twentieth century. The United States, Japan, the Marshall Islands, Kazakhstan, and more—who knows where next?

The Trinity site sits in what is now the White Sands Missile Range, one of two places in the continental United States with overflight restrictions that stretch all the way into space (the other is the White House). Like some kind of new shrine, the missile range opens its gates twice a year on a designated Saturday in spring and fall. It's a pilgrimage of sorts, and the traffic jams stretch for miles to permit thousands of atomic tourists to visit the site.

On an early spring Saturday, a yellow explosion of distant fusion over the eastern mountains heralds the day as I drive to the site. I proceed east into its piercing glare through the one-stoplight town of San Antonio, which still hosts the Owl Bar & Café that once fed the atomic scientists one of the best green chile cheeseburgers in the state. This road continues east through the scrubland to Roswell, where the aliens visited. It's dirt-poor here; the soil isn't much good for anything but ranching even when irrigation water is available.

As I pull off the highway and into the line of cars waiting to get through the White Sands Stallion guard gate, Richard Wagner's "Ride of the Valkyries" flickers onto my radio. I love the smell of fission in the morning. Locals hawk Trinitite—the special mineral fused from the dirt by that first blast—from minivans on the side of the road, just $20 for a tiny, still weakly radioactive pebble.

Radio towers bristle on the far hills, spring breakers and retirees in recreation vehicles add to the crowd. A dark brown calf watches, curious, as the long line of humans in cars inches forward, bisecting the range. The older cattle keep their heads down and graze. "It's like going to the Super Bowl," one of my fellow tourists jokes, and it's an hour-long stop-and-go traffic jam just to reach the Stallion Gate.

Cop car lights flash red, yellow, and blue at the gate. The police let us through in batches after a quick scan of a driver's license. A few protesters line the road outside the gate, waving skulls. Signs read SPEAKING UP FOR

THOSE SILENCED BY THE BOMBS and WE ARE THE TRINITY DOWNWIND-
ERS. All seven billion–plus of us are downwind of some nuclear bomb or
another now.

Judging by the animal crossing warning signs, imported oryx from
the Kalahari Desert of Africa continue to proliferate here in the Chihua-
hua Desert of North America, yet another sign of the human propensity
to reshuffle species. There are even hunting areas, at least four because
I pass Hunting Area 4 en route. The missile range covers more than 160
kilometers north to south, and the nuclear pilgrims are set loose in the
range like the oryx, and also herded, no doubt also like the oryx. A guard
in a white SUV and some orange cones block further progress, necessitating a turn left and heading back to the east and into the mountains. Just
as the ground begins to slope up, just past the remnants of the bunker like
some booby trap from *Mad Max* where some mad few watched the first
explosion, present-day soldiers guide the procession of cars into parking
spots just past a fence that bears the warning sign CAUTION RADIOACTIVE
MATERIALS. There's an ambulance at the ready.

By 9:15, only a few have finished the pilgrimage and already zoomed
back to Albuquerque or other civilized points. It's like a Burning Man for
old people, a guy in white wandering around and blowing a shofar. Ven-
dors hawk Day-Glo green DUCK AND COVER T-shirts, and the National
Park Service has set up a table full of goodies. A metalhead in a black
T-shirt that repeats Oppenheimer's famous first thought postblast "Now
I am become Death, the Destroyer of worlds" strolls by a family of four
with masks covering their mouths and noses. Grit, some of it kicked up
by the eight-person trams shuttling the elderly and infirm in and out,
lines my nose and throat. Trinity dust covers my shoes. My lifetime dose
of radiation is getting high thanks to visits like this one, tours of nuclear
reactors, all those long flights over the pole to China, and, most egre-
giously, all those dental X-rays. One sign of wealth in these Anthropocene days may be a bit more radioactivity than most.

Disaster tourists chat amiably about visiting Mount St. Helens and
being surprised by the recovery. There's not much to differentiate the
scrubland behind the fence from that of anywhere else on this range.
People stoop to look for Trinitite and, oddly enough seventy years later,
still find it, a spray of tiny green sandy pebbles here and there in the dirt.
(A large sign threatening that removing Trinitite is "theft of government

property and can result in fines and jail time" seems not to be a deterrent.) "I thought I'd be walking in a crater," another tourist observes. A self-proclaimed ex-Marine missing his front teeth wanders around with a slew of dosimeters and other gadgets to measure radiation. He claims to have still detected cesium over by Ground Zero, even though it has been through two half-lives' worth of decay by now.

Green Trinitite might be the source rock for the Anthropocene, along with the even rarer red varietal, pitch-black Atomsite from Nevada's long-lasting test site, and the brown-black fused agglomerate called Kharitonchik from the Soviet Union's test range called Semipalatinsk.

The journey embarked on with the Trinity test could make dead men of us all, and in far less time than it takes to cross that stretch of New Mexico desert. There is not much good news on the nuclear weapons front decades into this new epoch. Russia and the United States still maintain thousands of weapons on hair-trigger alert and plan to refurbish aging armaments. China continues to outfit more missiles, now with multiple warheads. Israel awaits Armageddon with its arsenal while India and Pakistan engage in a new arms race. The bizarre hermit kingdom known as the Democratic Republic of Korea churns out weapons-grade fissile material and has tested missiles both from land and sea. Meanwhile, the world watches to see if Iran will join this infamous club. If the former Persians do, another arms race may ensue as its wealthy neighbors and rivals like Saudi Arabia scramble to match the feat.

All humans are cyborgs, incorporating tools into our lives, whether it's as simple as fire or a hoe, or as complicated as an iPhone or a semiautomatic machine gun. But now we have created technology that exceeds us in its capacity to destroy. Atop rockets, these weapons could wipe out the world of civilization in moments. Of course, it is also those same rockets that could make us a multiplanet species if folks like Musk get their wish.

As I drove away from Trinity, the line of vehicles waiting to get in stretched nearly six kilometers. Will even more tourists flock to the site if it becomes ground zero of a new geologic epoch, one named after the species that made atomic weapons possible?

Only one planet bears life for sure, but there is a planet ruled by robots. That planet is Mars, which hosts two rovers crawling across its surface and five orbiters surveying the planet whole as I write this, allowing us to glimpse, say, the eerie beauty of sunset on an alien world. The surface bears the remnants of a dozen or so dead robots. That may not change in the near term, no matter how much Musk would like to die on Mars and add his organic matter to the regolith.

The robot planet is exclusively inhabited—if that word even applies—by metal animated with electricity generated from radioactive decay and radio signals from intelligent meat-sheathed skeletons back on Earth, just as drones may one day rule the skies and seas of this planet. It may stay that way for quite some time, more and more robots calling the Red Planet home, doing more and more tasks, ostensibly building a future for humans but perhaps simply building what will become a kingdom of what those same meat sacks call artificial intelligence, as if it was so different from our own thoughts.

Science is what made this possible, the patient application of the scientific method that is itself the largest collaborative project yet undertaken by humanity. From the early days of the Enlightenment in Europe to today, this method—observe the workings of the world, make a guess as to how it works, test that guess, observe some more—has accumulated knowledge that has enabled humans to live beyond one hundred years, to fission uranium atoms, and to send robots to Mars.

The future may be humans as gamers, controlling drones, robots, and replicants who will show us things we cannot believe, or see for ourselves thanks to the rigors of space travel and the hostile environment of space. In that case, we are already there. Some of the Mars robots boast sarcastic avatars on Twitter.

There is a bit of a new national space race going on, most of it focused on Mars. India and the United Arab Emirates aim to get there, as do the United States and the European Union. China thus far prefers to focus its efforts on the moon. Then there are the private companies attempting to do what nations have heretofore done: Boeing, Orbital Sciences, Lockheed. Rounding out this geopolitical rogues' gallery are the companies that can tap the deep purses of billionaires: flamboyant Richard Branson's Virgin Galactic, secretive Jeff Bezos's Blue Origin and, of course, Musk's SpaceX. The ultimate goals of Branson and Bezos remain opaque due to their obfuscating personalities, albeit at opposite ends of the spectrum. But Musk wants to seed migration into space, and his methods are clear:

reusable rockets to bring down cost. "So long as we continue to throw away rockets and spacecraft, we will never have true access to space," Musk has said.

Branson has much the same idea, though his SpaceShip Two is really just a glorified roller-coaster ride to the high atmosphere for the rich (or the financially negligent). Still, even that short trip may prove important, given that people see best when looking from the dark into the light. According to the astronauts, cosmonauts, taikonauts, and everyone else who has ever been to space, the view from up there really does change your perspective on down here.

The first camera to photograph the whole planet was on a satellite, prosaically named in the way of the American space program Lunar Orbiter 1. After a useful life of a little under a year, it crashed into the far side of the moon a few days before Halloween in 1966. As the historian Robert Poole observed, "Perhaps one day some lunar archaeologist will retrieve the remains of the first camera to photograph the Earth."

The rockets that brought that camera—and eventually men—to the moon derive from an arms race. Long before people were strapped to the nose of a rocket, bombs were, and the German scientists who rained V2s on London, like Wernher von Braun, became the forefathers of American spacefaring. And while the people haven't changed, the bombs have gotten bigger, stronger, more destructive.

Space is the pinnacle of human technical achievement, the conquest and inhabitation, even if for short periods of time, of an environment completely hostile to human life. Without the kind of CO_2 capture and storage needed to cleanse the atmosphere, human space flight will be impossible. In fact, without the full suite of incremental technology, from fossil fuels to polymers that recycle CO_2, life in space represents perhaps the hollowest of promises.

Being at the edge of human technical capability, space can go horribly wrong. The astronauts aboard the *Challenger* space shuttle felt and heard their rockets go silent after exploding. They then flew across the sky before plummeting to their deaths on impact with the ocean. For more than two minutes, those astronauts knew they were going to die, despite buckling their seat belts and inflating survival rafts. The same may happen to space tourists.

As Musk himself says, 99 percent of human effort should be focused on remedying Earth, hence Tesla's electric cars and SolarCity's photovoltaic build-out. But even 1 percent of global gross domestic product might be sufficient to build a self-sustaining city on Mars, enough to make humanity a multiplanetary species and lessen the odds of an asteroid doing to us what a similar bolide did to the big dinosaurs. That's just human life, and whatever microbes or animals we take with us, at least for starters. Not Life with a capital *L* as it flourishes on this planet.

Musk's argument is defensive in nature. Humanity faces threats that have reconfigured biodiversity in a geologic blink in the past. So spread our bets to more than one planet. "We want to be a spacefaring species, a multiplanet species, and ultimately out there exploring the stars," Musk says. "It's a much more exciting future than if we're forever confined to Earth until some eventual extinction event."

That statement contains the other argument for space: inspiration. Moving to Mars seems to inspire people to heroic effort—look at the employees of SpaceX. And we will never explore space if we wait for all the problems of Earth to be solved.

As some are fond of saying, the frontier never ended, it just moved offworld. If there is something to be explored, then some set of humans will explore it. In fact, the urge to explore may be part of life itself, like cells that grope toward food, proliferate, spread. To do otherwise seems somehow alien. To leave something unexplored seems nearly impossible for some of us to do, if not imagine. In these times, that urge is intensified by the potential for profit. Space will be explored and assayed until the profit is found, as long as there is time. Even in the absence of profit, exploration continues. To leave something unexploited also appears alien, though we do occasionally try with our parks and other conservation efforts.

Musk may want to die on Mars, but he has already almost died on Earth a few times, once after contracting malaria on his honeymoon in Africa. "Vacations will kill you," Musk told his biographer, Ashlee Vance. But he had another near-death experience in a car, the McLaren F1 his newfound wealth allowed him to possess long before Tesla.

Musk is clearly a car guy. After selling his first company—Zip2, an effort to get local businesses listed on a map and outfitted with rudimentary Web pages in the dawning days of the World Wide Web—to Compaq for more than $300 million, his first extravagance was to purchase one of the roughly one hundred McLaren F1s in the world.

But Musk is also the practical sort. So he turned that million-dollar sports car into an everyday vehicle. And once, on a trip up to Sand Hill Road where the venture capitalists lurk to seek more money for his next start-up, X.com, soon to be PayPal, he slammed down the accelerator to show what his expensive toy could do.

Turns out what it could do was not only accelerate into a ditch but also flip and spin like a Frisbee. The car was totaled and Musk was lucky to walk away. According to legend, he still made his VC meeting, hitching a ride. If so, it's no wonder that a failed rocket launch or hectoring from critics leaves him seemingly unfazed.

His enthusiasm for cars remained undimmed, though he reportedly took driving lessons in the wake of the crash. Not long after, he invested $10,000 in a start-up electric car company called Tesla. A few years later he would own the company outright. Musk's experience with fancy sports cars led him to push for more comfort from the Roadster and then, later, more zip from the potentially sedate sedan, the Model S. That impetus continues with the gull-wing doors for the Model X sport-utility vehicle, and the affordable luxury of the Model 3.

He's noted that he would like these electric cars to be powered cleanly, which means at a minimum nuclear power but also wind and the heat from rocks or, most preferably, electricity from sunshine. To that end, he talked his cousins Lyndon and Peter Rive into starting a solar company.

Solar has to be the future in Musk's mind because there's simply so much of it about, a gigawatt per square kilometer every day. Obviously not all—or even much—of that gets turned into electricity, but even a fraction could clean up the world's energy supply, at least during the day. At night, there are all those batteries Musk needs for his Teslas. The Gigafactory he's building in the Nevada desert can also pump out cells to store clean electricity for when it's needed.

Above all, there's his bid to get off the planet, inspired by reading science fiction in his youth in South Africa, a time that still can be heard in his uncanny English diction. He says he read all the time as a child, morning to night, a mix of fantasy, science fiction, even the encyclopedia. He found great comfort in books, by all accounts. "When I was a kid, I didn't know what I would be when I grew up," he says, much like all kids. "I didn't think anything great would happen at all."

Nevertheless, after failing to secure a job at Netscape, Musk had enough confidence to drop out of Stanford after enrolling in its PhD program for physics. The idea was to start an Internet directory. "I expected to fail," Musk says. The math agreed with him, since at least 90 percent of start-up companies fail.

Instead, he succeeded, with a few thousand dollars cobbled together from his own savings, his brother Kimbal's stash, a loan from his father, and a friend's money. They rented an office instead of an apartment, sleeping on the couch and showering at the local YMCA in Palo Alto. They had only one computer, Musk's, which meant the Zip2 website shut down at night when Musk was programming.

But in the go-go years of the dot-com boom, Zip2 was interesting enough to garner a payday. In 1999, four years later, Compaq bought Zip2 for $307 million; Musk's share was $22 million.

Musk moved on to attempt to revolutionize banking with the Internet, after observing during an early internship that bankers were not that bright. It became an effort that would end up in the PayPal system still used today. That earned Musk $180 million, but instead of retiring, he finally had the capital to do what he really wanted to do: change the world.

He returned to the fantasy world of his childhood and his science-fiction dreams of humans on Mars. Even the school Musk sends his five sons to reflects this obsession, a small school Musk himself started in 2014 and calls Ad Astra (to the stars), which is also the name of the magazine put out by the National Space Society. There they learn the purpose of a screwdriver or a wrench by taking apart an engine.

Back in 2001, he started up the Mars Oasis project with his old college roommate, a crackpot scheme to send a few seeds in a greenhouse to Mars to start making the first oxygen—as well as inspire another generation to reach for the stars. But it turns out it's not cheap, or easy, to get a greenhouse to Mars. In fact, the Russians won't sell you the rockets for it, and even trips aboard the rockets they send to space anyway are too expensive for such a stunt. Passion is fine, cold hard cash is better.

So instead, in 2002, Musk channeled that passion into the Space Exploration Technologies Corporation, or SpaceX. The goal was to provide a cheap way to the stars. In car terms, SpaceX wanted to replace the super-expensive McLarens (and Teslas) built by NASA and its usual contractors with a sturdy, reliable, cheap car just big enough to do the job, like a Toyota Camry or Honda Accord—the same way that Japanese automakers disrupted American car hegemony in the 1970s.

The probability of success was even lower for SpaceX than any of his previous companies. Ditto for the other companies he got involved with around that time: Tesla's electric cars and SolarCity's move to make photovoltaics cheap and easy for homeowners. But the potential pay-off—spacefaring civilization, clean air, sustainable energy—seemed large enough to justify any risk. By 2006, Musk was CEO of SpaceX and Tesla, and chairman of SolarCity.

Now Musk had the opportunity to watch all his dreams nearly die. Problems plagued Tesla's cars, SpaceX kept blowing up rockets instead of launching them into space, and money—easy money outside of Musk's own fortune—became hard to come by as the world entered the Great Recession. Musk says he came close to having something toward the end of 2008 he used to pooh-pooh: a nervous breakdown. He couldn't sleep as that fateful year drew to a close. "It was looking pretty grim," Musk recalls.

In fact, Musk had to borrow money to pay his rent after bailing out Tesla in 2008. His first divorce didn't help, especially as his divorce lawyers had to be involved in the negotiations with the federal government to secure a loan for the electric carmaker. But, in the end, Tesla got its outside money, closing on a fresh credit line in the last hour of the last day possible: 6 P.M. Pacific time on Christmas Eve 2008.

And just before Christmas, NASA too called with a lifeline for SpaceX—a $1.4 billion contract to ship supplies to the International Space Station. "I said, 'I love you,'" Musk recalls, "which I don't think is a normal response for NASA contracting officers." Given the number of companies NASA keeps in business, it may be.

Musk is not done. He tosses around transportation ideas—vacuum-tube hyperspeed trains—the way others speculate about what's for dinner, always adding that he will get to that in the future if time allows. "Not that I like failure, but if you only do things that are certain to succeed, you will only be doing very obvious things," he says.

In the meantime, there are those companies to run, a constant grind that involves adapting the basic idea—a cheap rocket, a cheap electric car, cheap solar—and refining it, improving it, and working hard and harder. As Musk says, a company is just a group of people gathered together to produce a product or service. And a company should exist only as long as that product or service is useful to customers. Musk would like to look back from his deathbed—On Mars? You never know—and think his actions had a positive impact on the world.

"You can only change the world for electric cars if people can afford to buy a great electric car," he says. "Affordable electric cars are not great, and the great electric cars are not affordable."

There is hope out there for a hero, someone who can save the world from all its problems. In the United States, for the moment, that hero may be Elon Musk. In a sense, we are living in a quest that he has devised, even if it isn't entirely original to him—electric cars and solar power to clean up this planet and rockets to spread life to another one.

That leaves him far too busy to meet with seekers like me. Or as his indefatigable Tesla PR pro Alexis Georgeson puts it, "Due to the dynamic nature of both Tesla and SpaceX, Elon is unable to accommodate interviews. However, he appreciates the interest."

A sky smeared with pink fades to white behind the buttes and slumping mountains. A dazzling burst in the high clouds heralds the sun until a blazing brightness tinged yellow rises past the San Andreas mountains to the east. This is our fusion reactor at a comfortable remove: the sun.

Spaceport America rises from the New Mexico desert in the middle of the Jornada del Muerto. The middle of nowhere is a good place to blow up things and crash things—not a lot to hit—but perhaps a bad place for a transit hub.

Space has become a kind of secular religion for some, a bit more respectable than Scientology. I head out on a tour of Richard Branson's folly on what Catholics call Good Friday, the day the Romans crucified Jesus. The signs on local fences say NO WOLVES, and I half expect to see a wolf similarly stretched out on hung barbed wire as a warning.

Branson is not the only rich man to find room in the New Mexico desert. Ted Turner is the largest local landowner, following in a tradition that dates back to royal patents granted by the Spanish king hundreds of years ago. Vultures riding the breeze and circling over the sagebrush are a more common sight than people. After all, the borough of Brooklyn where I live has twice as many people as the entire state of New Mexico. There are more cows than people. As the slogan goes: "I need my . . . Space."

Going to space is, in some sense, getting away from all the riffraff. Signs boast of a neighborhood METH WATCH along with NO TRESPASSING. There are lots of zeros in the bank account separating most folks from astronaut status.

The nearest "city" to Spaceport America is Truth or Consequences, a town that traded away its original designation as Hot Springs for a chance to host the game show of the same name on April Fool's Day 1950. At least the town and its "geothermic waters" are doing better than Engel, a ghost town that still appears on the maps even though no one lives there anymore. Little lizards and roadrunners scurry through the scrub, while grackles and mourning doves rest on the power lines.

Spaceport America proclaims itself the "world's first purpose-built commercial spaceport." It won't be the last—SpaceX is building another one near Brownsville, Texas, much to the displeasure of local residents, who resent the restrictions on their movement for safety reasons during future launches. There the buffer is the Gulf of Mexico, but in New Mexico it's the adjacent White Sands Missile Range, which keeps air traffic from passing over almost every day of the year. It's also the high desert, already a mile up in the sky, which is why Spaceport America likes to claim that "the first mile is free."

The buildings of Spaceport America glitter in the distance like a mirage, but they are all too real. There is no cell phone service, and everything necessary for a launch is brought in and then brought back out again. But there is infrastructure: roads, sewers, a facility to treat the dirtied water, power lines, fuel storage tanks, and enough water to put out two rockets on fire: 5 million liters of precious fluid. "You can never have enough water," my tour guide Gary Ramshaw says of life in New Mexico in general. He adds, "When you live in New Mexico, you have to change your favorite color from green to brown. Every shade of brown known to man out here."

The roads around the earthen berm that protects the starchitect-designed spaceport itself bear appropriate names, like Asteroid Beltway.

The spaceport, which looks uncannily like Mos Eisley from the desert planet Tatooine in *Star Wars*, looks out on a giant runway, 3,000 meters long, 60 meters wide, and 8 centimeters thick to handle more than 40 megapascals of force. With these buffers, the spaceway is just a bit under 5 kilometers long and plumbs a line 20 degrees off from due north and south.

It costs $250,000 to book your ticket to space on Virgin Galactic, and there's a waiting list for a spaceship that will hold two crew and six passengers per flight. That wait list probably has some overlap with the 90,000 reservations Pan Am once took for trips to the moon back in the late 1960s—a trip that never happened. A Virgin Galactic ticket won't even get you to real space, just 100 kilometers up in the sky to where the blackness of real space can be seen above and the curvature of Earth will become apparent below. Oh, and weightlessness, maybe for four minutes, maybe six, depending on the flight. It sounds to me like the most expensive roller coaster yet devised, and one that will just earn you your astronaut wings. On the other hand, it's a bargain compared to the $73 million per flight the Russians charge.

SpaceX has yet to offer tickets, but Musk's company is also testing in the New Mexico desert out of a couple of trailers with a satellite uplink. SpaceX's Grasshopper rocket aims to launch and land vertically on robotic legs, like something out of a Buck Rogers comic book from the early twentieth century.

Meanwhile, the U.S. government seems to be out of the space race for now, preferring to support Musk's SpaceX dream and others with tax dollars. But hope springs eternal, kind of like the hot springs in T and C. It's spas and *Star Trek*, and you can book a stay in the Apollo Room at the Rocket Inn.

Escape from Earth is perhaps the ultimate fantasy. Wherever we go in the universe—and that could be nowhere other than the moon where we've already been, briefly—we take Earth with us: Earth's gravity, Earth's atmosphere, the mentality that evolved on Earth. A space suit is just the smallest kind of human shell, one to keep us on Earth even as we travel to Mars, and one that would have to be replicated and expanded to make life there even possible. As Lucianne Walkowicz, an astrophysicist at the Adler Planetarium in Chicago, puts it, "The way a lot of private space industry talks about Mars, it's as though Mars is some sort of lifeboat for planet Earth, that if we were to screw up this planet, as we're well in the process of doing, unfortunately, Mars will be there to save us, and that it will be a backup to humanity."

The trouble is that if we don't know how to manage Earth's environment, we're not going to be able to figure out how to "terraform" Mars, a planet with wholly different characteristics from ours. "If we think we can change Mars and make it habitable, then we should be up to the challenge of keeping Earth habitable, and we should test those ideas here," Walkowicz says. We should think about Mars, she adds, as a place that can teach us about Earth. Musk would prefer to just repeatedly nuke Mars from space to put a lot of carbon dioxide into the Martian sky, fast. Or perhaps those old, patient geoengineers the cyanobacteria can help, given their success at transforming Earth and the ability to make use of what the Red Planet has to offer in terms of resources. Mars currently has the climate of Antarctica and the atmosphere of a vacuum chamber, exposing any human inhabitants to the radiation levels of an X-ray—all day, every day. That's a problem that would have to be remedied, fast, as would other medical challenges. At least Antarctica can provide a blueprint for how we might want to govern any Mars exploration—a treaty that governs international cooperation at some future research bases, even if those bases end up exclusively inhabited by robots.

My children and I like to imagine, and sometimes those musings include interstellar giants capable of devouring entire planets. The escape strategy in such an event is clear: colonizing, with rocket ships, Mars or a more distant world, like Io, even the sun itself. But, as always, there are the practicalities: A space suit for people that can withstand the heat and energy of the sun or the freezing cold of far-off (demoted) dwarf planet Pluto. A spaceship that can hold a fair number of people along with the water and the food they need, or at least the basics to make food at the press of a button from some yet-to-be-invented wonder machine. And underneath it all the need for a propulsion system that can get all that weight into space in a timely fashion and ensure human survival for both the duration of the flight and once colonization begins. But don't worry, my son, Desmond, is going to invent a supermetal that can do all this, or perhaps just Iron Man suits for everyone. My daughter, Beatrice, on the other hand, wants to build a giant bubble, perhaps made of water, a dream shared by some Sci Foo-ers.

Or perhaps you prefer to get your space advice from grown-ups. As Steven Chu, a physics Nobel laureate and former U.S. secretary of energy, has noted, the moon "is not a good place to live." Nor is Venus or Mars; in fact, no planet that astronomers have yet to discover is as fit for human habitation as Earth. That is perhaps obvious, but it is worth noting that cleanup will always remain our best option.

People are looking for a hero, and depending on ideological bent, that could be Elon Musk or Pope Francis. But human life in the Anthropocene is not so much heroic choices or efforts. It's billions of smaller choices and decisions, like Fan's quest for a better life for his daughter, replicated a billion or more times. Or a storm water garden that soaks in the rain before it floods into local waterways or overwhelms the sewers, bringing raw waste with it. This is ameliorating the engineering mistakes of the past and reengineering in a new, better direction, one that no doubt will need further correction in the future.

Like the decision to put solar panels on a roof, which is what finally brought me and the robot-like billionaire Elon Musk together.

The occasion was the market debut of SolarCity, an idea hatched from a road trip to the Silicon Valley pagan rite of summer known as Burning Man. "If Elon said, 'Get into a cheese factory,' I would get into a cheese factory," Lyndon Rive says of that long-ago trip to the desert.

Their mothers were twins, but it's the boys, the brothers and cousins, who remained close. Musk practically bounced into the air-conditioned room looking out over Times Square on a late summer day in Manhattan. He looked around like an expectant puppy for his chair, outfitted in the Silicon Valley uniform of a tech billionaire: a dark, expensive suit with a bright white shirt, open at the collar. Ties are for the millionaires.

"It didn't seem like there was enough entrepreneurial talent in the solar industry," Musk recalled of Lyndon and his brother Peter, now SolarCity's chief technology officer. "Since they're really talented, I thought solar could use their talents."

The idea became simple: Install solar panels on a home, then reap any federal tax credits or state incentives as well as the steady stream of payments from the homeowner. As the solar panels themselves become cheaper and cheaper, the constraint to growth becomes money—the capital to put more and more solar panels on more and more houses in those regions where it makes economic sense either because of sunshine (Arizona) or incentives (Massachusetts). "Arizona is the most hostile market," Lyndon Rive says.

"Ironically, it's pretty sunny over there," Musk jokes.

"The East Coast is a little more receptive," Rive adds. "In Arizona, it's not a super environmentally friendly state." On the other hand, Phoenicians and others in the Arizona desert, customers everywhere, are strongly attracted to the idea of energy independence, control of their own energy, even when the money saved versus utility rates is not that big.

Environmentally friendly reasons or not, the sunshine-generated electricity cuts down on the need to burn fossil fuels. "There's a giant fusion generator in the sky that runs for free, and we can catch the energy as it comes to Earth," Musk told me and the rest of the crowd in the Times Square headquarters of the NASDAQ MarketSite. "You can argue about when, but not if it will go overwhelmingly solar."

Given the number of homes in the world, the market is infinite, for the time being, especially if solar swells to become a major provider of electricity, even as demand for electricity also swells because of electric cars and the like. And, of course, Musk's Tesla will be happy to sell you an electric car or a battery pack to eliminate disruptions like a cloud passing over your solar panels, or sunset, churned out by the "ludicrously large" factory it is building in the Nevada desert. In a cleaner world, a world where the line of soot marking the dawn of the Anthropocene is very thin, the rest of our electricity needs could be provided by the wind, dams that use falling water to spin turbines, and heat harvested from hot rocks deep underground. The fission of atoms could play a role too, but "I assume not a huge number of nuclear reactors because most people are not super comfortable with a nuclear reactor being built in their backyard," Musk says.

Plus, the solar panels of the future may make a house look better, making the energy benefits more of a bonus, a world where a child's drawing of a home no longer includes an archaic chimney for burning wood but instead something gleaming and modern on the roof or in the windows that perhaps sizzle with electricity. Just as a Tesla electric car is just a better car than the alternatives, as long as you have the money, a solar system will have to be better—and cheaper. But, as my daughter asks, how would

Santa get in? There are cultural barriers to a better future that must be overcome to get there, and it is a future that will be very different from place to place.

As it stands, the CO_2 capacity of the planet's atmosphere and oceans is being used up, like dump space, without charging anyone for it. That zero cost—which also encompasses all the hacking coughs, asthma, heart disease, and other health issues that come from breathing air fouled by burning fossil fuels—distorts the energy market, making fossil fuels look cheaper than they actually are. "We have an error in the market information system, that's what it is," Musk explains. "The best thing would be a carbon tax, to price carbon instead of it being a price of zero."

"It seems like a commonsense thing to do to base tax collection on things that are more likely to be bad than things that are good, like jobs," he adds, then chuckles, bemused perhaps by the foibles of the rest of humanity, without access to his iron logic. "I know common sense doesn't always play, I'm just telling you."

But Musk is very, very serious about these issues, committing his personal fortune to making a solar-powered, electrically driven future that turns humanity into a multiplanet species. "It's been obvious to me since college, we have to have sustainable production and consumption of energy. The primary source of primary energy will be the sun. The sun powers all of Earth already. It powers the system of precipitation, the entire ecosystem is solar powered. We're talking about capturing a teeny-weeny bit of that and using it to power human civilization," he says. "This just seems, like, super obvious. What other answer could there be?"

The goal is not disruption, in the argot of Silicon Valley business; the goal is to make things better than they are.

I waited patiently for my chance to get Musk alone, but there were plenty of other seekers, sitting, waiting in the audience as he talked on of other uncompetitive markets, like the military-industrial complex keeping SpaceX from competing for U.S. Air Force contracts, the vagaries of working for NASA, and the fact that Manhattan does not have enough roof space to generate all the electricity needed to run that world capital, though perhaps with PV-impregnated windows . . . "In places like Manhattan, you can still generate 10 or 20 percent of the power from rooftop solar and that would still be good," he offers. "We might as well do that."

The cost of solar keeps getting cheaper, and it will continue to do so as the technology improves and, perhaps more important, as the processes for making it get bigger and more efficient. "We're not saying delay your solar installation," Musk says. "Just think of all the money you can save between now and then."

And this is not just an American thing. Even Saudi Arabia, the kingdom that oil built, is trying to go solar. "I've been meeting with senior people in the Chinese government," Musk adds. "I think China is going to do a carbon tax."

"Eventually, they'll have a revolt if they don't," Lyndon Rive interjects, and this may finally make solar electricity rather than just solar hot water accessible there.

The show over, Musk and Rive join me and a few other reporters in a back room, where I ask Lyndon to tell me what working around Elon is like.

"I've spent thirty years competing with Elon," he says, laughing. "You do not want to compete with Elon." Even if it's just a game of Dungeons & Dragons.

"We've known each other for as long as we've been conscious," Musk adds, extolling the virtues of his cousin.

But Musk does fear something: artificial intelligence. "AI is something that is clearly happening much faster than most people realize," he says, a hint of concern creeping into his voice as he dreams out loud about the future. "Potentially great benefits and some very scary negatives, so we need to be super careful with that."

The negative consequence that worries Musk is the extinction of humanity when the artificial intelligence realizes that it no longer needs us, or the planet does not, a kind of superecologist, perhaps, ridding the world of the invasive species behind a gathering mass extinction so that the rest of life can continue unimpeded. Similarly, "the rate at which we transition to sustainable energy matters," Musk notes, a bit of didacticism steeling the timbre of his South African accent. He notes that anyone who does not crave a more sustainable future must be insane. "It matters whether we achieve sustainable clean energy sooner rather than later. We'll do much less damage."

So I ask him, as close to one-on-one as we're going to get: What are your future plans? Do you want to die on Mars? "Preferably not on impact and if I can be sure SpaceX will be all right." He thinks for a moment.

"There's an electric jet I'd love to pursue, but I need to stay focused on Tesla and SpaceX and stay reasonably sane. Maybe at some point, I'd like to do the jet thing, electrification of boats and, really, all transportation, except for rockets ironically."

With that, Musk is done, called away from dreaming of future Hyper-loops as public transport on Mars to more important present business. Perhaps Musk's greatest role is to show that most of us need bigger dreams. We need to recapture a sense of actively changing the world for the better, of striving toward a brighter future. There's an outbreak of clapping from my fellow journalists, which he seems surprised by and acknowledges with an awkward bow of his head as he bounces out of the room and off to his next appointment. Whether Musk ultimately succeeds or fails, he has already changed the game, like any half-decent Dungeon Master. We can protect Earth. Be the planet's stewards. Intelligence allows us to adapt and invent. Let's use those skills to fix what is broken. Leave the planet better than we found it. Musk, at least, seems to be trying. He is one of the titans of the modern age, another kind of robot perhaps to add to the collection on Mars and a hero eerily predicted by the Nazi father of rocketry and the U.S. space program, von Braun, who in his odd science-fiction book *Project Mars* had the planet ruled by ten technocrats whose chief bore the title "Elon."

No billionaire or technocrat is going to save us, whether your preferred model is Musk, his friend Larry Page, hard at work on that Elon-scaring AI, or Gina Rinehart, the billionaire whose Hancock Prospecting is attempting to wrest as much out of the Earth as is possible, environmental impacts be damned. Nor will it be the old school financial titans like info-billionaire Michael Bloomberg and his C40 group's effort to reduce greenhouse gas emissions in cities, or the hedge fund manager Tom Steyer with his focus on political action to combat climate change. Not even billionaires from the rest of the world, like China's solar power mogul Li Hejun or India's Ratan Tata. The world of capitalism would like to think that money flows to those with the best idea, but money actually flows to where it can grow.

There are other contenders for world savers, perhaps Pope Francis, who hopes to transform the relationship between the world's more than a billion adherents of the Catholic faith and the world itself. As Francis (and perhaps a coterie of ghostwriters including cardinals and other advisers) writes in his *Laudato Si'* encyclical: "Humanity has entered a new era in which our technical prowess has brought us to a crossroads. We are the beneficiaries of two centuries of enormous waves of change: steam engines, railways, the telegraph, electricity, automobiles, airplanes, chemical industries, modern medicine, information technology and, more recently, the digital revolution, robotics, biotechnologies and nanotechnologies. It is right to rejoice in these advances and to be excited by the immense possibilities which they continue to open up before us, for 'science and technology are wonderful products of a God-given human creativity.' "

Or, more succinctly: "Technology has remedied countless evils which used to harm and limit human beings," he writes. "Who can deny the beauty of an aircraft or a skyscraper?"

In fact, Francis thinks technologists like Musk and his teams of engineers, as well as all those who went before them, deserve gratitude and appreciation, especially as they help with the transition to a more sustainable civilization.

On the other hand, that technology can be misused: "It must also be recognized that nuclear energy, biotechnology, information technology, knowledge of our DNA, and many other abilities which we have acquired have given us tremendous power. More precisely, they have given those with the knowledge and, especially the economic resources to use them an impressive dominance over the whole of humanity and the entire world. Never has humanity had such power over itself, yet nothing ensures that it will be used wisely, particularly when we consider how it is currently being used."

We, or at least a small part of the world's human we, have access to technologies that can kill millions, or perhaps reduce civilization to rubble and shadows. "In whose hands does all this power lie or will it eventually end up? It is extremely risky for a small part of humanity to have it."

That is equally true for the pope himself. Ultimately it is what his faithful do that matters more than what he writes, says, or does, even with all that concentrated, dictatorial, totalitarian power. Self-restraint, care for others, all the things that a good Christian is required to do, do not always happen. Priests molest children and the church covers it up. Women are demoted by tradition. Even in the part of the world without

God, people put their faith in impersonal institutions and systems, networks of experts functioning in tandem to produce outcomes few can foresee and perhaps none individually desire. Thus we live in the world we do.

Scattered across North America, hiding in deep launch tubes carved out of rock, slumber the U.S. nuclear missiles known as the Minuteman. These are the heirs of missiles that promised total annihilation, superweapons, stenciled with names like Peacekeeper, but also Snark or Hound Dog. As the AI in the movie *War Games* said of the great game these missiles were designed to play: "The only winning move is not to play."

Relic missile sites outside Tucson are one of the few places in Arizona where you are *not* allowed to bring a gun. Climb down into the bunker and travel back in time to where it's always 1963 behind the three-ton blast door. Here nuclear warheads tipped rockets exactly like those used in the Gemini space program, waiting for a world war that might last all of ninety minutes. A total war of horrifying efficiency as cities on at least two continents are erased from the map and the world is plunged into radioactive years without summer.

Moon suits are needed here, too, in order to handle the toxic fuel of these apocalyptic rockets. Four soldiers sat down here, waiting for orders that, thankfully, never came, orders to end the world. The bunkers smell of machine oil and dust and stored up to thirty days of food and water, though only fifteen of air (which seems an oversight). These nuclear jobs remain the most boring in the world until, all of a sudden and without warning, they become the most important and challenging in the world.

The Minuteman missiles, along with nuclear-armed submarines and bombers, still wait, arrayed around the world by the United States, matched by weapons in China, Russia, India, Pakistan, and elsewhere. Death would fall from the sky, a ballistic free fall from space, and then a light three times brighter than the sun to wipe out 2,500 square kilometers of Earth. As the slogan goes: "Peace is never fully won. It is only kept moment to moment."

Polemically, this present time is another outbreak of the long-standing fight between those who think the world is irredeemably broken and those who think it's getting better all the time. The truth is both are right (or wrong) as people muddle through as best we can. The worry is not so much that the system cannot grow but that it cannot change. Wisdom,

restraint, humility are everywhere in short supply, but there is no short-
age of will to tackle problems like extinction or poverty, and hubris is
perhaps on the rise. A rising tide can lift all of life if the right choices are
made. Or, through ignorance or inertia, the powerful can leave a legacy
of rack and ruin. We need new rituals and new observances that will help
us confront the Anthropocene and our own anthropocentrism.

The kids in my son's kindergarten class note: Wouldn't it be cool to
control the world? To make it stop snowing or even rain chocolate? Why
not? But there can be no doubt that those who would intervene in the
weather itself would ultimately see that used as a weapon too. To be able
to give the rain is also the ability to withhold it.

Tweaking, adjusting or, in words more familiar to the Musks of the
world, "rapid iteration" is how humans muddle through, making something better, building on what came before, improving, breaking free
from the trap of history. Not so much undoing the mistakes of the past as
acknowledging and repairing, as best as possible, those mistakes, whether
reviving foolishly discarded traditions or inventing new ones. Not far
from where SpaceX's factory now stands were once the wetlands of the
Los Angeles River, a meandering flow channeled into a concrete drainage
ditch like so many other urban rivers, streams, and creeks too numerous
to count. Where once bears feasted on salmon now there are humans
dreaming of Mars, and California has no bears except on its state flag.

We could end up with a kind of self-referential playground, a world
refracted through human needs, wants, and whims. Humility and com-
passion are required now more than ever. And a willingness to refrain, to
let a vacant lot takes its own course without human improvement.

I don't presume to speak for anyone else in this. There is already too
much speaking for others. We will need all the different ideas we can get
to make the Anthropocene last. But I do believe there is no hero in this
story. The hero is all of us, as is the villain. The hero is none of us.

No one, whether Elon Musk or Pope Francis, can solve all the challenges we face, certainly not alone. Having been to enough global negotiations around climate change, I know that no government—even of the
people, by the people, for the people—is going to solve these problems
either, the Paris Agreement notwithstanding. It's going to take the efforts
of all seven billion of us, just as it takes all those homeowners to work
with Elon Musk, not to mention the Herculean efforts of Lyndon Rive

and the employees of SolarCity and all the other solar companies to make a future powered by sunshine a reality.

There's good news. Renewable, cleaner energy sources are everywhere. No country or company or person has a lock on sunlight or the wind or Earth's heat beneath our feet. That can translate into political power, given the potential numbers of those who could benefit, much as the automobile created a decentralized world of freedom of travel driven from the bottom up. That's the political potential of the renewables revolution.

On the other hand, such political power can be diffuse, compared to the concentrated might of a fossil fuel company or a petrostate. How these battles—like the war between solar homeowners and electric utilities in the United States—are won or lost will, slowly but surely, map out the road we will take into the future. Change comes from extremists pulling the great mass of the disinterested in one direction or another, changing the impossible to the possible, then the plausible before the probable, and finally the done. We must put our faith in each other, rather than heroes or gods, and our best efforts into our shared institutions and systems.

I once asked Musk how he survives all this effort. What drives him? "I wonder that myself, " Musk said, falling into a reverie before continuing on with a chuckle and a grin. "It does end up being a lot. I wake up and think: 'Where am I?'" In other words, not even Elon has a master plan, or truly knows why he does what he does. The ultimate mystery remains human nature.

Space travel, to Mars and ultimately even beyond, is not a luxury for a species that wants to last on geologic timescales. As the saying goes, if the dinosaurs had a space program they might still exist today. The wise use of our technology to create sustainability might possibly enable a long-lasting civilization. Human flexibility allows us to adapt to changing circumstances over and over again, the move down from the trees, off the savannah, across the world, adapting to encroaching ice and then its retreat.

Human nature may not change much; the basics are the same. Unless we really go in for enhancements, and perhaps end up permanently empowered by an artificial intelligence wired into our brains and enabling endless modifications of the physical self, like extra senses.

All while untold billions find such resources unobtainable and perhaps still seek clean water. In other words, not much will have changed. In a democracy, we get the future we deserve.

What else won't change? Old buildings will still be here, perhaps retrofitted. Old cities will still live as new cities thrive. Old ideas will thrive too, whether Catholicism and Islam or some resurgent paganism. Just as the Russian Orthodox Church sprang back from decades of repression under the Soviets, nature worship may yet return like a spring bloom.

All the things we'd need to live in space or on Mars—resilience, adaptability, and at least a short-term plan—are what have gotten us this far as a species. They are also what we're going to need here on Earth, our perfect planetary home. If, in some far future, we do expand into space, it will require us to have our home in order, so that we transform future homes into it.

It could be worse. Getting to Mars or dreaming of resurrecting a dead bird is at least a fruitful use of human intelligence. Author William Gibson has described subcultures as "a sort of unconscious R&D, exploring alternate societal strategies." We're going to need all the subcultures, all the lingos and languages, everything we have ever learned to craft an enduring Anthropocene.

Once we get past the loss of nature as we have known it, an important and meaningful period of mourning for what we have done, then we can perhaps begin to have a hopeful conversation about what kind of future we want to make.

We will need to build a broad, even catholic "we." A we that encompasses hovel world and hotel world. It is a dubious assumption that humanity constitutes one species, capable of solidarity, as the humanities and history have taught us over and over again. Or, as Pope Francis puts it, "Human beings, while capable of the worst, are also capable of rising above themselves, choosing again what is good and making a new start."

We are going to need all that the humanities can teach us about ourselves to build this we, and to get comfortable with the uncertainty that attends all complex systems, whether the sensitivity of the climate, increasing levels of carbon dioxide, or the breaking point from dissension and rancor to civil war. We will also need all we can learn from the rest of life on Earth about redundancies, adaptability, flexible systems for control, and vastly decentralized decision making.

Making an ever greater we will be hard, but it is the essential work of politics, community, and family. To take more care than to care about control. How big a we can we make, even if a global we is a kind of Zeno's paradox, always just out of reach?

And what comes next? A cosmoscene where we leave Earth behind as a nature preserve and spread into the solar system, the galaxy, and beyond? Even in such a future, humanity would remain Earthbound, tied to this planet, perhaps incapable of survival without it. We think we are the end, the Anthropocene name even implies it, but we are not and likely never will be, even if we make it far into the future.

On the cosmic scale, humanity seems insignificant, as is Earth itself, even the sun, the solar system, the Milky Way galaxy. Even here at home, humanity is insignificant in the grand sweep of geologic time. As psychoanalyst Sigmund Freud observed, "Great revolutions in the history of science have but one common and ironic feature: they knock human arrogance off one pedestal after another of our previous conviction about our own self-importance."

But in the moment, right here, right now, we are also supremely significant, and that is what the Anthropocene acknowledges. We have world-changing impacts and the chance to engage in planetary protection for the first time. We represent a new intelligence emerging on this planet at this point in deep time, just as the existence of metals required billions of years of stars being born, living, and dying in spectacular explosions that spread these new atoms throughout the universe, seeding the future Earth with the elements that would make modern technology possible.

The pictures from space reinforce this idea, whether Apollo 8 or the latest from the Suomi satellite. The planet is huge, and yet it is just a "pale blue dot . . . a mote of dust suspended in a sunbeam" from even as near as Neptune, as astronomer Carl Sagan famously remarked. The pictures remind this self-important species that we are even smaller, invisible in many ways, and that the planet endures, no matter what we do. At the same time, this special place is fragile, and we can wreck it for ourselves like some tantruming teenager who cuts off his own nose to spite his face.

Modernity often seems to be the world permanently at risk, apocalypse. Modernity is also a world shaped by human choices, and the risks they entail, whether it is future people with enhanced genetics or world-straddling and nuclear-armed AI born from self-driving cars. As the joke goes, it remains debatable whether there's intelligent life in the universe, or even here on Earth. Artificial intelligence and cyborgs could change that.

We're too busily embarked on a course of befouling our only home. Humanity lives in the midst of a great acceleration, our powers and numbers multiplying. Biology has been reset, species moved around by truck, ship, and airplane, humanity as asteroid. The life that has adapted to a semiregular cycle of shifts from Ice Age to summery span and back again; moving with the ice and temperatures will now be taken out of that fluctuating comfort zone and cast into a prehistoric, almost premammal, warmer world of extreme weather and sunken coasts. The Arctic will be transformed, including an ice cap melted away, if that warming continues, which will surely leave a permanent mark.

One of the beliefs that make this transformation possible, even tolerable, is that there's somewhere else to go if we break this planet. This seems to be an extension of the idea that technology will save us, as it has in the past. But being saved by technology always has costs, and those costs are too easily glossed over or ignored in this thinking that is both wishful and necessary. There is danger in trusting powers once reserved for deities to fickle people. If we are become as gods, some of us are the evil ones.

Geologic time is at least a way to get everyone on the same clock, synchronizing our timekeeping devices of whatever kind. Time is the fourth dimension, and getting us all in synchrony is a feat like ordering all the gaseous molecules in the atmosphere. Too many are trapped in the past with attempts to re-create the Islamic caliphate, fighting the truth of climate change, or, for that matter, trying to bring back the Pleistocene. It's a mistake to think life was better in the past. As geologists know, the only constant through time is change. As long as we're here we will always have the chance to make better or worse choices. That may be the ultimate utility of this Anthropocene epoch idea.

We need a science conspiracy directed toward solutions, like synthetic biology for preserving ferrets or drones to monitor the lungs of Earth, rather than simply discovery. What are real people's problems? Identify those and solve them. Can some form of capturing the abundant energy poured forth by the sun usher us into an age of ubiquitous energy from devices, perhaps called the Stereocene as chemist Ugo Bardi has suggested, the next epoch in a long-lasting Anthropozoic? We're all world building now, either through growing the Internet hive mind with our tweets or the choice to grow this or that plant in a garden, keep that tree or meandering brook, or level that land and channelize that runoff. Building the broadest possible we means using the best knowledge of all of us, like pairing traditional, soil-preserving farming practices with crops tweaked genetically to withstand extreme weather.

First there was the runner, the marathoner of old who dropped dead when he delivered the news, then the man on horseback, or in a swift boat, like the ones that united old and new worlds. Now, like the telegraph before it, the Internet connects the minds of people into one seething cauldron of ideas and opinions. Further shrinking of the globe may yet come via satellites or some new network pervading the planet itself and everything on it. Space and time have already been annihilated by instantaneous thought and response.

The Internet has grown into a global nervous system, sending alerts from the far parts of civilization's body to centers where there's the wherewithal to deal with them, whether it is ProMED-mail, a LISTSERV that propagates health news and alerted the world to the first outbreaks of bird flu and Ebola, the global outrage spawned by the gruesome death of Cecil the Lion in Zimbabwe, or the efforts to finally stamp out the lingering vestiges of slavery.

But life is not a one or zero, no matter how much some may try to convince us otherwise. The Anthropocene is not a yes or no proposition; it is a world of increasing gray, where some choices are slightly better in some ways, slightly worse in others—and we will have to find a way to choose. Uncertainty is the soul of science. We need to promote this quieter, humbler approach, full of doubt rather than artificial certainties. What are the odds that the growing panoply of scientists and thinkers of today have it all right? Zero. We are going to have to get comfortable with complexity, diversity, all the things the human mind hates. Nature is indifferent, only humans care, whether for nature or themselves.

The Anthropocene is about the connections between earth and sky, sea and land, what you buy and what you believe. Most important, it's about connections among us. The Anthropocene applies to individuals, families, tribes, and societies. We can't claim that such social structures just happen; we build them. We may even be entering a kind of Hyper-

Anthropocene, our impacts and the changes they cause escalating and accelerating. We don't want to live in such an Oligarchocene I suspect. Our collective dreams will make for a better or a worse future.

The Anthropocene may be the wrong name or even bad geology, but it is the right idea. The Anthropocene forces a shock of recognition of the sheer scale of human impact and the need to respond in a new way. It doesn't matter what the geologists decide in the end, people need to craft a new time.

In the course of writing this book I have learned one thing for sure: Despair is an ideology we cannot afford. Optimism without action may be immoral, but fatalism inevitably spawns apathy and inaction. One cannot be an optimist in the sense that all is as it should be, but it's also far too late to be a pessimist. Hopeful, relentless work to make the world better remains the best path forward.

Whether this is the Anthropocene, the Holocene, or some other name, we are living through the end of geologic time, synchronizing with history. This is an immutable fact, especially since the burning of all those fossil fuels will smudge out our ability to date the past through fossil carbon, obscuring where we came from for all the future. It's not epochs we need to worry about, it's the end of civilization or humanity itself.

This is our time, now.

Acknowledgments

To properly acknowledge and thank everyone who helped in the genesis and fruition of this book would take a tome in itself, including a numberless legion of friendly, informed folks from the interwebs such as my fellow members of the Anthropocene book club, and journalists of all stripes, as well as scientists and the just plain interested. So I say thank you to one and all, you know who you are, or should.

But there are those who must be thanked by name for extraordinary service in the making of *The Unnatural World*, and that list starts with my own family: my beloved Shii Ann, who excites my every day and tolerated with good grace the countless intrusions of this particular "rock job" on her life. Of course, this book would not have been possible without the inspiration of my daughter, Beatrice, and son, Desmond, who give me great hope for the future and this new people's epoch. I also thank my mother, Elizabeth Burleigh, and my mother-in-law, Lili Yeh, for timely assistance in matters great and small, not least free babysitting, as well as the Biello, Burleigh, Dumbleton, Huang, and Pan clans, all of whom helped even if they didn't always know it.

I would be remiss if I overlooked thanking my work family at *Scientific American*, including (but not limited to) Mike Battaglia, Lee Billings, Mariette DiChristina, Mark Fischetti, Larry Greenemeier, Robin Lloyd, John Matson, Ryan Reid, Kate Wong, and Phil Yam, who all inspired and enabled this book over many years. That work family thanks extends to my extended *SA* clan: Katherine Harmon, Marguerite Holloway, Christopher Mims, JR Minkel, George Musser, Christie Nicholson, John Rennie, Nikhil Swaminathan, and many more. My reporting was improved by time spent long ago working with Mark Nicholls, now a freelance editor and journalist in the UK, whom you really ought to hire, and my storytelling has been improved (if not perfected) by time spent at Disney Feature Animation in my callow youth, so thanks to all my former colleagues there as well. The list of friends I've made during my employment years is a long one, and it has been one of the great privileges of my life to work with all of you.

Then there are those whose hard work made this published work possible: Will Lippincott, whose friendship and tenacity I do not deserve but am thankful for, as well as Ethan Bassoff, who brilliantly guided this wayward writer. I also thank everyone at Scribner, from Paul Whitlatch, who first saw the potential in this nascent book, to Colin Harrison, who helped realize that potential, as well as Sarah Goldberg, who helped make possible all the little things that prove so big in the end. Cynthia Merman is perhaps my most careful reader and I thank her for both improving this book and catching at least a few of my errors and typos. For timely advice on publishing and book writing, thanks also to Steve Levine, Eric Roston, and Jeffrey Rotter.

This book would not be what it is without the help of my Super Secret Reader Committee whom I will dare to name here: Ross Andersen, Gal Beckerman, Tim Biello, Katie Fehrenbacher, Chris Heiser, Mara Hvistendahl, Barry Johnson, John Matson, Ed Moore, and Robin Lloyd. I'm also grateful to Marguerite Holloway (again) for book-writing inspiration and Samuel Freedman for book-writing praxis. Both ensured my time learning at Columbia University was well spent, which is an educational list that must also include Tim Harper, who is the harshest editor I may ever face. Special shout-out to the Dark Side, who continue to inspire me every day, as well as my old friends from Wesleyan and life in StL, LA, and NYC.

Thank you to everyone who appears in the pages of this book for being so generous with your time and thinking: Victor Smetacek, Jan Zalasiewicz, Mark Williams, Stephen Gienow, Erle Ellis, Dana Boswell, Jonathan Dandois, Ariane de Bremond, Ben Novak, Beth Shapiro, Stewart Brand, Ryan Phelan, George Church, Matthias Stiller, Hung Chih-Ming, Robert Lanza, Peter Heintzman, Dov Sax, Cun "Angela" Yanfang, Arun Majumdar, Lai Hun Suen, Feng Yu, Li Shiyang, Fan Changwei, "Frank," Yan Hui, Ma Jun, Zhang Guobao, Peter Kelemen, Ah-Hyung "Alissa" Park, Peter Eisenberger, Klaus Lackner, Allen Wright, Wally Broecker, Paul Olsen, Elon Musk, Lyndon Rive, Alexis Georgeson, Gary Ramshaw, and Lucianne Walkowicz. Some who didn't appear in the final text also deserve thanks: Andy Chaikin, Karthik Ram, Megan Palmer, Nina DiPrimio, David Grinspoon, Kim Stanley Robinson, Charles Mann, Ted Merendino, Ninad Bondre, and Owen Gaffney, as well as all the participants of the Symposium on the Longevity of Human Civilization in 2013 and the 2014 Anthropocene Synthesis Workshop. That, of course, in no way implicates any of these fine people in any of the conclusions or presentations of fact, which are mine and mine alone, along with any and all errors.

And my final thanks is to you, for reading and, no doubt, contributing to a better Anthropocene for us all.

Selected Notes
and Further Reading

The Overview

Chaikin, Andrew. *A Man on the Moon: The Voyages of the Apollo Astronauts*. New York: Penguin Books, 2007.

Crutzen, Paul J. "Geology of Mankind." *Nature* 415 (January 3, 2002), doi: 10.1038/415023a.

Poole, Robert. *Earthrise: How Man First Saw the Earth*. New Haven: Yale University Press, 2008.

Schwägerl, Christian. *The Anthropocene: The Human Era and How It Shapes Our Planet*. Santa Fe: Synergetic Press, 2014.

Waters, Colin N. et al. "The Anthropocene Is Functionally and Stratigraphically Distinct from the Holocene." *Science* 351, no. 6269 (January 8, 2016): 137, doi: 10.1126/science.aad2622.

White, Frank. *The Overview Effect: Space Exploration and Human Evolution*. Boston: Houghton Mifflin, 1987.

Chapter 1: Iron Rules

Roberts, Callum. *The Unnatural History of the Sea*. Washington, D.C.: Island Press, 2007.

Smetacek, Victor et al. "Deep Carbon Export from a Southern Ocean Iron-Fertilized Diatom Bloom." *Nature* 487 (July 19, 2012): 313–19, doi: 10.1038/nature11229.

Stager, Curt. *Deep Future: The Next 100,000 Years of Life on Earth.* New York: St. Martin's Press, 2011.

Walker, Gabrielle. *Antarctica: An Intimate Portrait of a Mysterious Continent.* New York: Houghton Mifflin Harcourt, 2013.

Chapter 2: Written in Stone

Beresniewicz, Irena. *Hands Across the Precipice: Encounters in Siberia and Asia 1940–1942.* Wroclaw, Poland: Wydawnictwo APIS, 2010.

Finney, S. C. "The 'Anthropocene' as a Ratified Unit in the ICS International Chronostratigraphic Chart: Fundamental Issues That Must Be Addressed by the Task Group." Geological Society, London, Special Publications, October 24, 2013, doi: 10.1144/SP395.9.

Mann, Charles C. *1493: Uncovering the New World Columbus Created.* New York: Knopf, 2011.

Seielstad, George. *Dawn of the Anthropocene: Humanity's Defining Moment.* Alexandria, VA: American Geosciences Institute, 2012.

Thomas, William L. Jr., Carl O. Sauer, Marston Bates, and Lewis Mumford. *Man's Role in Changing the Face of the Earth.* Chicago: University of Chicago Press, 1956.

Vince, Gaia. *Adventures in the Anthropocene: A Journey to the Heart of the Planet We Made.* Minneapolis: Milkweed Editions, 2014.

Zalasiewicz, Jan. *The Earth After Us: What Legacy Will Humans Leave in the Rocks?* New York: Oxford University Press, 2009.

———. "The Epoch of Humans." *Nature Geoscience* 6 (January 2013): 8–9.

———. *Planet in a Pebble: A Journey into Earth's Deep History.* New York: Oxford University Press, 2010.

Chapter 3: Ground Work

Dandois, Jonathan P., and Erle C. Ellis. "High Spatial Resolution Three-Dimensional Mapping of Vegetation Spectral Dynamics Using Computer Vision." *Remote Sensing of Environment* 136 (September 2013): 259–76, doi: 10.1016/j.rse.2013.04.005.

Ellis, Erle C., et al. "Anthropogenic Transformation of the Biomes, 1700 to 2000." *Global Ecology and Biogeography* 19, no. 5 (September 2010), doi: 10.1111/j.1466-8238.2010.00540.x.

Marris, Emma. *Rambunctious Garden: Saving Nature in a Post-Wild World.* New York: Bloomsbury, 2011.

Nash, Roderick Frazier. *Wilderness and the American Mind*. 4th ed. New Haven: Yale University Press, 2001.

Quammen, David. *The Song of the Dodo: Island Biogeography in an Age of Extinction*. New York: Scribner, 1997.

Wapner, Paul. *Living Through the End of Nature: The Future of American Environmentalism*. Cambridge, MA: MIT Press, 2013.

Wuerthner, George, Eileen Crist, and Tom Butler. *Keeping the Wild: Against the Domestication of Earth*. Sausalito, CA: Foundation for Deep Ecology, 2014.

Chapter 4: Big Death

Carlin, Norman, et al. "How to Permit Your Mammoth: Some Legal Implications of 'De-Extinction.'" *Stanford Environmental Law Journal* 33 (January 2014).

Church, George, and Ed Regis. *Regenesis: How Synthetic Biology Will Reinvent Nature and Ourselves*. New York: Basic Books, 2012.

Fuller, Errol. *The Passenger Pigeon*. Princeton: Princeton University Press, 2014.

Kolbert, Elizabeth. *The Sixth Extinction: An Unnatural History*. New York: Henry Holt, 2014.

Mooallem, Jon. *Wild Ones: A Sometimes Dismaying, Weirdly Reassuring Story About Looking at People Looking at Animals in America*. New York: Penguin Press, 2013.

Pääbo, Svante. *Neanderthal Man: In Search of Lost Genomes*. New York: Basic Books, 2014.

Powell, William. "The American Chestnut's Genetic Rebirth." *Scientific American*, March 2014.

Shapiro, Beth. *How to Clone a Mammoth: The Science of De-Extinction*. Princeton: Princeton University Press, 2015.

Chapter 5: The People's Epoch

Brooke, John L. *Climate Change and the Course of Global History: A Rough Journey*. New York: Cambridge University Press, 2014.

Cohen, Joel. *How Many People Can Earth Support?* New York: Norton, 1995.

Hvistendahl, Mara. *Unnatural Selection: Choosing Boys over Girls, and the Consequences of a World Full of Men*. New York: PublicAffairs, 2011.

Pearce, Fred. *The Coming Population Crash: And Our Planet's Surprising Future.* Boston: Beacon Press, 2010.

Tattersall, Ian. *Masters of the Planet: The Search for Our Human Origins.* New York: St. Martin's Press, 2012.

Chapter 6: City Folks

Boo, Katherine. *Behind the Beautiful Forevers: Life, Death and Hope in a Mumbai Undercity.* New York: Random House, 2014.

Hodder, Ian. *The Leopard's Tale: Revealing the Mysteries of Çatalhöyük.* New York: Thames & Hudson, 2006.

Mehta, Suketu. *Maximum City: Bombay Lost and Found.* New York: Vintage, 2005.

Nagle, Robin. *Picking Up: On the Streets and Behind the Trucks with the Sanitation Workers of New York City.* New York: Farrar, Straus and Giroux, 2013.

Saunders, Doug. *Arrival City: How the Largest Migration in History Is Reshaping Our World.* New York: Vintage, 2011.

Watts, Jonathan. *When a Billion Chinese Jump: How China Will Save Mankind—Or Destroy It.* New York: Scribner, 2010.

Chapter 7: The Long Thaw

Alley, Richard. *Earth: The Operator's Manual.* New York: Norton, 2011.

The Keeling Curve, Scripps Institution of Oceanography, https://scripps.ucsd.edu/programs/keelingcurve/.

Kelemen, Peter B., and Jürg Matter. "In Situ Carbonation of Peridotite for CO_2 Storage." *Proceedings of the National Academy of Sciences of the United States of America* 105, no. 45 (November 3, 2009): 17295–300, doi: 10.1073/pnas.080574105.

Kolbert, Elizabeth. *Field Notes from a Catastrophe.* New York: Bloomsbury, 2006.

Lackner, Klaus S. "Washing Carbon Out of the Air." *Scientific American,* June 2010.

McPhee, John. *Annals of the Former World.* New York: Farrar, Straus and Giroux, 2000.

Chapter 8: The Final Frontier

Arthur, W. Brian. *The Nature of Technology: What It Is and How It Evolves.* New York: Free Press, 2011.

Billings, Lee. *Five Billion Years of Solitude: The Search for Life Among the Stars*. New York: Current, 2013.

Elias, George Henry. *Breakout into Space: Mission for a Generation*. New York: William Morrow, 1990.

Finney, Ben R., and Eric M. Jones. *Interstellar Migration and the Human Experience*. Berkeley: University of California Press, 1985.

Rhodes, Richard. *The Making of the Atomic Bomb*. New York: Simon & Schuster, 1986.

Index

acid rain, 86, 198
Africa, 60, 121, 140–45. *See also* South
    Africa
agriculture, 4, 44, 54, 72, 84, 89, 96
  as altering Earth, 44, 48
  birth of cities and, 171, 172
  carbon dioxide release and, 215
  in China, 82–85, 160, 190, 193
  famine and, 12, 82, 84
  hectares in farming, 136
  "long summer" climate and, 156
  loss of trees and, 80, 81
  nitrogen fertilizer, 83, 85, 207–8
  organic farming, 89, 167
  soil conservation, 82
  techno-optimism and, 85
  water needs, 139
Ahad, Shaheda, 11–12
air pollution, 5, 31, 33, 136, 178, 194,
    203, 204. *See also* carbon dioxide
  air quality index, 171
  cars and, 171, 187, 191, 198–99, 233
  in China, 170–71, 176, 178, 182, 184,
    187, 191, 192, 193, 194, 196–99
  detecting, 79, 96
  dust, 187, 223–24

health and, 143, 176, 187, 228, 255
  in India, 174
  industrialization and, 144, 174, 176,
    179, 187
  as invisible, 210–11
  lack of desire to change, 214
  U.S., 176
Alfred Wegener Institute (AWI), 12, 24,
    26, 27, 31, 34
algae, 12, 14, 16, 20, 22, 32, 37, 47, 215, 223
algae blooms, 184, 190
aluminum, 6, 169, 202, 236
Alvarez, Luis, 3–4
Always, Rachel (Carson), 201
Amazon rain forest, 15, 23, 81, 90, 93
Amazon River, 37
Amyris company, 145
Anders, William "Bill," 1–2
Anderson, Chris, 105, 106
Aniene River, 94
animals, 4, 8, 22, 53, 97–130. *See also*
    de-extinction; extinctions; *specific*
    animals
  DNA of rare, frozen, 120
  domesticated, 61, 84, 101
  endangered, 124, 128, 129, 134

animals (*cont.*)
  human shaping of survival, 137–38
  invasive species, 122
  "outbreak species," 112
  of the Pleistocene, 71, 103, 107
  rare, in zoos, 112
  reshuffling of locations, 241
  rewilding, 120–21
  species vs. hybrids, 119
Antarctica, 11, 27, 29, 35, 37, 56, 204
Antarctic Circumpolar Current, 33
anthromes, 71, 72, 73, 77, 83, 133, 136
Anthropocene, 6, 7, 8, 57, 82, 127, 155–56
  as age of fire, 59, 93
  alternative names for, 43, 93, 173
  anthropocentrism and, 55, 81, 260
  arguments about, 42–43, 58, 59
  changed climate as sign of, 222
  China as crucible for, 84
  choices and decisions for, 253
  cities and, 165–200, 262
  coining of term, 42, 52
  as "doomsayer" or opportunity, 59
  Ellis on, 82, 84, 92, 96
  ending of, 66
  the Enlightenment and, 64
  in fiction, film, and song, 53
  fossil fuels as driving, 47
  garbage as marker of, 57–58
  geological proof of, 39–66
  global government for, 91
  goal for, 60
  history of idea, 53
  homogeneity and, 62
  human-animal relationship, 97–130
  idea of, 92
  improving humanity and, 131–61
  index fossil for, 40, 61
  longevity of civilization and, 153
  as a meme, 52–53
  new human perspective for, 60
  ocean alteration and, 11–38
  personal doses of radiation and, 241
  religion and, 262
  restoring ecosystems and, 35

  science-based policies and, 64
  shaping human thinking, 92
  Snyder on, 86
  space and space travel, 232–66
  start date, 4, 44–46, 65, 71, 84, 239
  state of the Earth survey and, 90
  thermal maximum, 231
  what comes after, 62
  what it is, 2, 265–66
  what won't change, 261–62
Anthropocene Working Group, 42, 43,
      58, 63
Anthropogene, 55
Anthropozoic, 54, 63, 65, 66, 265
Apia, Samoa, 199
Apollo 8 mission, 1–2, 3, 263
Arctic, 2, 45, 62, 101, 117, 119, 127, 206, 264
Arctic National Wildlife Refuge, 73
Arctic Ocean, 214
Arendal, Norway, 166
Argo floats, 37
Arizona State University Center for
      Negative Carbon Emissions, 219
Armstrong, Neil, 14
ARPA-E, 146–48, 209
Arrhenius, Svante, 221, 228
artificial intelligence (AI), 66, 96, 152,
      256, 257, 259, 261, 264
Ashoka of India, Emperor, 48
Asian black bear, 134, 158–59
Asian elephant, 107, 115
asteroid impacts, 3, 5, 100, 152, 245
Athens, Greece, 173
atmosphere, 6, 7, 20, 23, 81, 139. *See also*
      air pollution; carbon dioxide
  CO_2 buildup, 13, 29, 206, 215, 220–21, 228
  CO_2 comes out of, three ways, 208
  composition of, 4, 13, 22, 205
  geoengineering of, 224–25
  human intervention in, 23, 36, 207–8
  ice cores to determine prehistoric, 204
  methane in, 44, 84, 207
  nitrogen extraction, 85, 207–8
  "optical depth," 81
  ozone layer, 224

rock layers, information in, 204–5
weight of dry vs. wet, 26, 26n
Audubon, John James, 111, 113
Audubon Society, 99
Aurangazeb, Mughal emperor, 175
Australia, 101, 103, 153–54

baiji (river dolphin), 112, 120, 149
band-tailed pigeons, 118, 124, 128
Ban Ki-moon, 141
Bardi, Ugo, 264–65
Barro Colorado island, 90
beaver, 113, 120
Beijing, China, 174, 176–79, 181, 199
  air pollution, 176, 187, 191, 194, 196,
    197–98
  consumerism, 196
  growth, 198
  highways of, 191
  illumination at night, 178, 196
  political leadership in, 197
  water supply, 189
Benxi, China, 181
Beresniewicz family, 49–51
Bezos, Jeff, 243
Biello, Beatrice, 252, 254
Biello, Desmond, 252, 260
biofuel companies, 146
biogas, 157–58, 160
biome (ecosystem), 72, 73, 74, 77
Biosphere 2, 218
birds, 15, 27, 97, 118, 121. See also
    specific birds
  biggest killers of, 61, 111
  extinctions, 97–100, 105, 110–14
  theropods as ancestors of, 101–2
bison, 101, 121–22, 124
black-footed ferrets, 127
black-handed spider monkey, 137–38
Blakney, Raymond B., 165
Bloomberg, Michael, 257
Blue Origin, 243
Bodélé Depression, 223–24
Boeing Company, 243
Boltwood, Bertram, 55

Borman, Frank, 1, 2
Bosch, Carl, 208
Boswell, Dana, 75–77, 80, 95
Bradbury, Ray, 64
Brand, Stewart, 104, 108, 109, 116, 127
Branson, Richard, 243, 249–51
Brazil, 4
"Bread from the Sea" (Davidson), 12
Bremen, 25–26
Bremerhaven, 13, 20, 24–26, 29
Brilliance company, China, 180–81
British Columbia, 30
British Geological Survey, 41
Broecker, Wallace, 218, 219, 221
Bronx Zoo, 121–22
bucardo (Iberian mountain goat), 115
Buffon, Comte de, 53–54
Bukowski, Charles, 39
Burning Man, 253
Burroughs, Edgar Rice, 224–25
Bush, George W., 183
butterflies, extinct, 105

California, 260
Cambodia, 80
Cambrian boundary, 57
camels, 103
carbonate rock, 201, 204
carbon dioxide, 4, 13. See also air pollution
  agriculture and, 215
  amount to be stored, 213
  atmospheric buildup of, 13, 29, 206,
    207, 215, 220–21, 228
  capturing, 203, 208, 209, 211–20,
    228–29, 230
  carbon footprint, American, 136
  carbon neutral cities, 165–66, 182,
    193, 199, 200
  cars and, 155, 180–81, 187, 191,
    198–99, 221
  CCS projects, 211, 212
  cement-making as source of, 212
  climate shift and, 206
  coal-to-gas and, 189
  deforestation and, 207, 221

carbon dioxide (*cont.*)
  duckweed to capture, 213–14
  evolution and, 65
  fossil fuel burning and, 4, 13, 16, 18,
      26, 29, 43, 65, 66, 140, 155, 166,
      189, 203, 204, 206, 207, 208, 212,
      214, 221–22, 228, 230
  Global Thermostat, 212
  global warming and, 201–31
  as invisible, 210–11, 223
  iron fertilization experiments, 17, 18,
      19–20, 28, 29, 32–37
  Keith's extracting machine, 23, 34
  krill to capture, 37
  Lackner's artificial tree, 209, 210,
      213, 216–17
  length of time to dial back
      atmospheric levels, 65
  mantle rock to store, 203, 227
  modern rise, start of, 207
  Oman's mountains and, 201–4, 214,
      229
  percentage of atmosphere, 205
  permanent changes made by, 65
  photosynthesis to capture, 13, 209,
      220, 223
  plankton to capture, 13–14, 37, 203,
      214
  rocks that store, 202–3
  scale of problem, 223, 229
  Southern Ocean as sink, 28
  storage of extracted, 201–4, 214,
      219–20, 226, 227, 228, 229, 231
  trees to capture, 18, 74, 95
  turning into useful products, 209
  urgency of reduction, 225
Carboniferous Age, 47, 142, 231
carbon tax, 255, 256
caribou, 101, 126
cars, 193, 233
  air pollution and, 187, 191, 198–99
  China and, 178, 179, 180–81, 190–91,
      193, 198–99
  CO_2 buildup and, 155, 180–81, 187,
      191, 198–99, 221

  number built yearly, 213
  Tesla Motors and electric, 232–39
Carson, Rachel, 201
Carter, Jimmy, 183
Çatalhöyük, 171, 172
cats, domestic, 61–62, 75
Cat's Cradle (Vonnegut), 23
Cattle Decapitation (band), 53
cave bear, 101
Ceauşescu, Nicolae, 122
Cenozoic era, 55
cesium 137, 44–45
chalk, 215
chickens, 118, 126
Chile, burning of Patagonia, 59
China. *See also specific cities*
  agriculture in, 82–85, 190, 193
  air quality, 168, 170–71, 178, 182,
      184, 187, 191, 192, 193, 194, 198
  American lifestyle adopted by, 177
  antiquity of writing system, 154
  carbon neutral plans, 199, 200
  carbon tax, 256
  cars and, 178, 179, 180–81, 190–91,
      193, 198–99
  circular economy and, 199–200
  cities of, 173–99
  clean energy, 863 Program, 147
  cleanup efforts, 194
  coal use, 180, 194
  computers in, 161
  consumerism in, 196, 199
  contrasts of, 157
  corruption in, 188, 192, 197
  cuisine, 157
  Cultural Revolution, 165, 167
  Dandois in, 77
  deforestation, 132–33
  driving in, 190–91
  ecocities, 181–82, 198
  economic growth, 169, 176, 198, 200
  education in, 169
  elephants in, 86, 168
  Ellis in, 77, 82–83
  English fluency in, 170, 200

environmentalism in, 131–35, 171
ginkgo tree in, 168
globalization and, 170
highway building, 174
housing construction, 166, 174, 178, 181, 187, 194, 197, 198
hukou system, 168
as human-dominated world, 84, 86
hydropower and dams, 148–51, 187
industrialization and pollution, 152, 178–79, 194
leaders of, 188, 197
moon travel and, 243
Mount Tai, 193
natural gas as fuel, 198
Naxi minority group, 132
nuclear power, 187
oldest alcoholic drink, 172
one-child policy, 132, 133
Petrochemical Research Institute, 190
photovoltaic production, 183
poplar tree in, 168
population density, 173
poverty in, 161, 176, 193
rainmaking and 2008 Olympics, 23
religious revival in, 199
Rust Belt, 179–80
scientism in, 187–88
sewage problem, 84–85
Shandong Province, 170, 199–200
Singles Day, 196
smoking in, 184, 193–94, 195
soil conservation, 82–83, 85
solar energy, 148, 157, 183
South-North Water Diversion, 189
steel production, 181
taxes in, 185
Three Gorges Dam, 149–51
Three-North Shelterbelt Forest Program, 187
transportation, 191–93
urbanization of, 178
war on pollution, 198, 199
water pollution, 189, 190
"water-stressed" cities, 189

wind power, 180, 187
wind turbine production, 176–77, 180
women in, 133, 159–60, 169
xiao jie, 197
Yunnan Province, 131–35, 158, 167
Chinese yew tree, 134
Cholula, Mexico, 173
Chongqing, China, 178–79
Chu, Steven, 147, 148, 253
Church, George, 107, 115, 127, 152, 158
circular economy, 166–71, 182, 199–200
cities, 4, 8, 154, 165–200, 262. *See also specific cities*
  air pollution, 170–71, 174
  carbon neutral, 165–66, 169, 184, 193, 199
  circular economy, 166–71, 182, 199–200
  clean energy and, 174
  disease and vanished, 140
  ecocities, 181–82, 198
  environmental impact of, 173–74
  illumination of, at night, 178, 196–97
  megacities, 178
  oldest cities, 172–73
  origins of, 171–72
  population, 173
  poverty in, 174, 175
  power laws for, 151
  religion and, 171, 175, 176
  resiliency of, 171
  rivers and, 260
  sewage problem, 188–90
  strata of trash in, 186
  sustainable practices and, 60
  transportation, 191–93
  waste problem, 186–87
  water pollution problem, 190
  "water-stressed," 189
Clements, Frederic, 72
Cliffs of Dover, England, 214
climate change, 4, 6, 32, 44, 48, 65, 66, 142, 145, 155, 184, 260–61, 264. *See also* global warming; Ice Age
  adaptation vs. opposition, 223
  amount of CO_2 to shift climate, 206

climate change (*cont.*)
~~CO₂ and global warming, 201–31~~
  computer models, 203
  drought, 223–24
  extinction of graptolites and, 39–40
  extinctions and, 127
  halting, 208–11
  intervention, problems with, 228
  lack of action, 222
  lack of will to change, 190
  "long summer," 70, 84, 102, 156
  new climate arrival, 222
  Paris Agreement, 220, 223, 229,
    260
  plants reshuffling and, 62
  Steyer and, 257
Climos, 18
cod, 15, 62
coffee, 137–38
coltan ore, 151
Columbian Exchange, 62
Columbia University, 218, 226, 227
  Kelemen and colleagues, 209–10
  Lamont-Doherty Earth Observatory,
    227–28
  Mudd Building, 216
  Schermerhorn Hall, 203
Columbus, Christopher, 62
Coming Population Crash, The (Pearce),
  137
commerce, innovation and, 52
common rock pigeon, 114
Congo, 80–81, 90, 136, 151
Consolations to Polybius (Seneca), 155
consumerism, 137–39, 175, 191, 199
continental drift, 27
Convention on Biological Diversity
  (CBD), 17, 19–20
Copenhagen, Denmark, 145, 166
coral reefs, 14, 16, 31, 32, 48
Corso di Geologica (Stoppani), 54
Costa Rica, 89, 90, 95, 166
cotton, 139
Crosby, Alfred, 62
Crutzen, Paul, 42, 52, 53

Cun (Angela) Yanfang, 131–35, 151,
  158–61
Cuvier, Baron, 53
cyanobacteria, 5, 12, 21, 22, 252

Damascus, Syria, 172–73
Dandois, Jonathan, 77–79, 81–82, 87, 95
DARPA, 146–47
Darwin, Charles, 41, 114
Davidson, Bill, 12
Dawkins, Richard, 52
Da Yu, 150
de Bremond, Ariane, 90–91
Deccan Plateau, India, 175, 227
Deepwater Horizon, 229
de-extinction, 104–19, 124–28
  candidates for, 118, 123, 127, 128, 234
  chimeras and, 119, 126
  cloning and, 115, 126
  CRISPR-Cas9, 115
  methods for, 114–15
  passenger pigeon, 107–14, 118–19,
    127
deforestation, 4, 5, 15, 70, 72, 80–81, 96,
  138, 207, 221
Delhi, India, 173
Democene, 153
Denmark, 147, 193, 209, 211
Devil's Bed, 35
dimethyl sulfide (smell of the sea), 20
dinosaurs, 100, 101–2, 110, 118
dire wolves, 71
DNA
  ancient, 100, 102, 112, 114, 117, 119
  Church's experiments, 107
  cloning and, 115
  contamination of, 109–10, 116
  dinosaur, 110
  heterochromatin, 116
  mammoth, 106
  Neanderthal genome, 102
  oldest on record, 107
  paleogenomics and, 103
  of rare animals, frozen, 120
  sample collection, 117, 124–25

source code for life, 152
stem cells and, 115
UCSC ancient DNA lab, 110
of viruses, 117
dodo, 99, 120
Doggerland, 26, 35, 117
Dolomites, Italy, 214
Dongtan, China, 181–82, 184
Donora, Pennsylvania, 176
drones, 69–70, 73, 78–79, 81–82, 87–88,
    94–96
    monitoring with, 70, 79–82, 94–95
    uses, 76, 78, 96
Dube, Pauline, 59
duckweed, 213–14
Durban, South Africa, 144

Earth, 4–8, 13, 14, 17, 21–22. See also
    atmosphere; oceans; specific issues
    in 1968, 3–4
    age of, 55
    agriculture and altering of, 72
    Archaean Earth, 21
    civilization on, at risk, 63
    core samples, 225–28
    current state, sample needed, 90
    "earth system governance," 59, 60, 91
    ending of, 66
    glacial ice and, 203
    human global responsibility, 6, 7
    humans dooming query, 63
    keeping habitable, 252
    "limits to growth," 59
    mantle of, 201–3, 226
    monitoring and mapping, 70–96
    most densely populated regions, 173
    Nagoya targets for conservation, 96
    optimal human population for, 137
    pollution and orbit wobbles, 203
    as seen from space, 1–2, 3, 263
    wars and reshaping of, 51–52
Earth After Us, The (Zalasiewicz), 41,
    46, 49, 58
Earthrise (photo), 3
Ebbehøj, Søren Lyng, 211

Ebola, 140, 142–43
ecocities, 181–82
ecology, 69–70, 89, 91–93
    aggressive intervention, 92, 94
    New York City and, 93
    norm for some humans, 93
    practicing in the modern era, 91
    rewilding and, 120–21
Ecosynth, 76, 79, 87
ecosystem. See biome
Edison, Thomas, 218
education, 157–61
    Chinese proverb, 161
    energy needs, 158
    of girls, 160, 169
Ehrenfeld, David, 105
Eisenberger, Peter, 209, 211–13, 230
elephants, 86, 106, 107, 115, 121, 129, 168
Ellis, Erle, 70–77, 82–85, 87–96, 104,
    237–38
Ellis, Ryan, 90–91
energy, 140–51, 187, 188. See also
    specific types
    American consumption, 136
    amount used yearly, 230
    ARPA-E and, 146–48
    clean energy, 145, 147, 152, 166–67,
        174, 261
    geologists and locating sources, 63
    Germany's goal of renewable, 25
    Kenya as green superpower, 142
    Khosla's black swan, 145–46
    for light, 139
    sustainability and, 141, 255
    Tennessee Valley Authority, 149
    United States as leader in, 148
    waste burned for, 187
    wasted, for lighting cities, 174–75
energy poverty, 140–45, 146
Enlightenment, 64, 92, 243
Eno, Brian, 108
environmentalism, 3, 5
    aggressive intervention, 92
    carbon neutral cities, 165–66, 182
    in China, 134–35, 158–59, 160, 194

environmentalism (*cont.*)
 conservation, 48, 54, 70, 73, 77, 92, 113, 121
 Earthrise (photo) and, 3
 green power, 142
 Nagoya targets for conservation, 96
 rewilding, 92
 role of hunters, 121–22
Eocene epoch, 213–14
Eremocene, 130
ETC Group, 17
Ethiopian wolf, 115
European Union, 148, 243
evolution, 5, 65, 86
extinctions, 15, 16, 44, 45, 53, 54, 61, 62, 91, 97–130, 155. *See also* de-extinction; *specific animals*
 causes, 44, 100, 104, 106, 111–12, 118, 120, 126–27
 coming of sixth mass extinction, 104
 mass extinction, Permian period, 45
 mystery of, 102
 number of, 101, 120
 preventing, 127
 speeding up of, 61

Facebook, 96
"fail fast" theory, 233
famine, 12, 82, 84, 85
Fan Changwei, 165–71, 184, 193–94, 195, 198, 199, 200, 253
Fang Wang, 169
Farmers of 40 Centuries, or Permanent Agriculture in China, Korea and Japan (King), 82
Feng Yu, 158, 167
Fengyuan, 169
Field & Stream, 122
Finney, Stan, 42–43
fish. *See also specific types*
 collapsed ecosystems and, 15, 62
 coral reefs and, 16, 32
 extinctions, 14, 15
 fishing, impact of, 15–16, 62

humans as evolutionary force on, 32
 rebound, 30
Fitzgerald, F. Scott, 72
food
 American obesity and, 192
 cooking, 142–43, 159, 228
 hectares in agriculture, 136
 impact on the Earth, 7
 lack of, 121, 135, 136
 meat eating and pasture, 136
 organic food, 199
 population growth and, 136–37
 smokeless cookstoves, 143–44
 unequal distribution of, 136
food chain, 12, 14, 29
forests. *See also* deforestation; trees
 benefits of, 80
 carbon dioxide capture, 18, 74, 95
 Central American, cutting of, 138
 conservation, 73, 160–61
 Global Forest Watch, 80, 81
 inventorying trees, 75
 monitoring, 70, 77, 78, 81–82, 88, 90, 94–95
 most at risk, 80–81
 primeval European forest, 73
 rain forests, 90
 reforestation, 74, 80, 93, 95, 122, 161
 temperate broadleaf forest, 72–73
 tropical, understanding of, 89–90
40 Million Salmon Can't Be Wrong (George), 37
fossil fuels
 acidifying of oceans and, 29
 alteration of the Earth and, 47
 American consumption of, 136
 amount burned, 228
 atmosphere composition and, 13, 18 (*see also* carbon dioxide)
 boom town of Williston, 98–99
 burning, CO_2, and global warming, 4, 13, 16, 18, 26, 29, 43, 65, 66, 140, 155, 166, 189, 203, 204, 206, 207, 208, 212, 214, 221–22, 228, 230
 burning, weather changes and, 20

for cars, 142, 155, 178
China, India, and increasing use, 230
clean coal, 211–12, 213
cleaning up, 228
coal, 47, 142, 144, 148, 149, 180, 187, 188, 194, 198, 210, 227, 228
coal, CO_2 emissions per second, 231
coal-to-gas mistake, 189
dangers known since 1950s, 221
earthquakes and extraction of, 61
fracking, 210
geology and, 47
increased use, 140
indoor air quality and health, 143, 167
industry, denial of problem, 230
natural gas, 193, 198
real cost of, 255
soot and, 45, 65
wasteful habits and, 210
fossils, 48, 61
graptolites, 39–40, 49
index fossils, 40, 48–49, 61
technofossils, 40, 46, 56–57, 88
Fourier, Charles, 160
Francis, Pope, 257–59
Laudato Si' encyclical, 258
"Frank" (Chinese official), 184–85, 188, 191–92, 193, 194, 195, 199
Franklin, Rosalind, 158
Freud, Sigmund, 263
future, 261–66
empowered women and, 160
human adaptation and, 261–62
human-nature link, 155
Musk and, 238
stories of a better world, 154
as unforeseen, 155
visions of, as wrong, 156

Gabriel, Sigmar, 17–18
Gaia (Lovelock), 41
Galápagos, 17, 18, 19
Gambia, 141–42
garbage, 57–58, 186–87
Genesis, 2–3

geoengineering, 17, 18, 23, 28, 30, 31, 34, 60, 224–25, 234. See also specific projects
Geological Society of London, 42, 43
geology, 39–66
age of the Earth and, 55
Anthropocene and, 40–46, 56
anthroturbation and, 56
chalk layer beneath Europe, 215
at Columbia University, 209
core samples, Palisades, 225–28
creating an Anthropozoic and, 65
diabase, 226, 227
fossil fuels and, 47
"golden spike," 40, 42
index fossils, 39–40
Oman, Earth's mantle in, 201–2
Orbis spike, 44
rock layers, information in, 204–5
soot as marker of climate change, 65
stratigraphy, 40, 42, 46, 57
what it does, 47
George, Russ, 18–19, 30, 37, 224
Georgeson, Alexis, 249
Germany, 25, 213
giant beaver, 71, 103, 104
giant kangaroo, 44, 101, 104
giant sloths, 44, 71, 103
giant snapper, 14
giant wombat, 101
Gibbard, Philip, 43
Gibson, William, 262
Gienow, Stephen, 69, 71, 79, 86–87, 94–95
Global Forest Watch, 80, 81
Global Research Technologies, 218–19
Global Thermostat, 212
global warming, 86, 201–31
bleaching of coral reefs and, 16
burning fossil fuels and, 4, 13, 16, 18, 26, 29, 43, 65, 66, 140, 155, 166, 189, 203, 204, 206, 207, 208, 212, 214, 221–22, 228, 230
meltdowns and, 33, 37, 56
as new normal, 222

global warming (*cont.*)
~~peridotite to reduce, 203–4~~
    predicted temperature rise, 221–22
    rising sea levels and, 56, 222
    sulfuric acid proposal for, 224
    Venus and, 214, 224
    volcanic eruptions to cool, 224
    wet bulb temperature, 222
Göbekli Tepe, 171
Goldberg, Dave, 209
golden monkey, 133, 135, 160
Goldwind factory, China, 176–77, 180
Google
    Earth, 2
    Earth Engine, 81
    Glasses, 153
    Project Tango, 96
    RE<C program, 148
    Sci Foo un-conference, 151–53, 252
graptolites, 39–40, 49
Great Acceleration, 45, 58
great auk, 105
Great Barrier Reef, 32
Great Gatsby, The (Fitzgerald), 72
Green, Richard "Ed," 102, 107, 114
Greenberg, Joel, 109
greenhouse gases, 13, 37. *See also* carbon
        dioxide
    C40 group, 257
    carbon footprint, American, 136
    carbon neutral cities, 165–66
    global warming and, 206
    hydrochlorofluorocarbons, 224
    rice farming and methane, 44
    Southern Ocean and, 28
    sulfur hexafluoride, 224
    Venus and, 214, 224
    Virgin Earth Challenge, 215
    wealthy vs. poorest humans, 136
Greenland, 56, 204
Greenpeace, 19, 178–79
GreenSea Venture, 18
"green tides," 31
ground sloths, 44
grouper, 15

Guangzhou, China, 199
~~Guinea, 81~~
Gulf of Mexico, 30, 226, 229, 250

Haber, Fritz, 45, 85, 208
Haida people, 30, 37
Haig, Susan, 126
Haldor Topsøe company, 212, 230
Hamilton, Clive, 59–60
Hancock Prospecting, 257
Hansen, John, 212, 230
Harari, Yuval Noah, 139
Harvard University, 92, 107, 195, 200
    ancient DNA lab at, 110
Hawaii, 32, 55
Hawaiian petrel, 15
Hayes, Dennis, 227
heath hen, 121
Heintzman, Peter, 116, 118, 119
Higgins' Eye pearly mussel, 106
Hodder, Ian, 172
Holmes, Sir Arthur, 55
Holocene, 4, 43, 44, 52, 54, 55, 266
    "long summer" climate, 70, 84, 207
Homo denisova, 47–48
Homo neanderthalis, 47–48
Homo sapiens. See also humanity
    appearance of, 172
    arrival in North America, 101
    atmosphere composition and, 220–21
    climate change, handling, 222
    disease as big enemy of, 140
    extinctions and, 44, 47–48, 100, 104,
        111
    genomes of, 116
    as invasive species, 122
    killing of megafauna, 104
    manipulations of habitat, 104
    movement of the species, 51, 103
    other hominids breeding with, 119
    population of, 45, 48
    as rulers of Earth, 4, 22–23
    shaping of their environment, 93
    wet bulb temperature, 222
Honolulu, 187

Horner, Jack, 118
horses, 101, 107, 114
hotel world, 93, 139, 155, 156, 174, 262
houbara (desert bustard), 118, 121
Hoyle, Fred, 2
Huaneng company, 198
Huangbaiyu, China, 181
Hu Jintao, 188, 197
humanity, 131–61. *See also specific issues*
  ability to change, 156
  consumerism of, 137–39
  cultivating kindness, 155
  education, 158
  energy use, 140–51
  fire, relationship with, 143
  knowledge of, 153–54, 156
  life expectancy, 142
  longevity of human civilization, 153
  mental architecture and new
    problems, 155
  population growth, 135, 136–37, 158
  storytelling, 153–54
  sustainability and, 131–35, 141
human nature, 155, 156
Humboldt, Alexander von, 54
Hung Chih-Ming, 113, 114
hurricanes, controlling, 23
Huxley, Thomas Henry, 41
hypsometer, 76

Ice Age, 39, 55, 102, 113, 154
Immelt, Jeffrey, 146
India, 11–12, 129, 174–76, 230, 243
  Kerala, Land of the Coconuts, 176
Indian Ocean, 15
Indian Point Energy Center, 138
industrialization, 173, 177–81, 194, 219
  CO$_2$ release and, 207, 212
  pollution and, 144, 152, 174, 176,
    178–79, 187,194
insects, 22, 102, 119
International Geosphere-Biosphere
    Programme, 52, 58
International Third Geological Congress
    (1885), 54

International Union for Conservation of
    Nature, 129
Internet, 52, 98, 105, 147, 232, 237, 265
  Musk and, 247
Inuit people, 62
Iraq, 51
Irish elk, 122
iron, 22, 27, 28
iron fertilization experiments, 17, 18,
    19–20, 28, 29, 30, 32–37
Iron Man (film), 232

Jakarta, Indonesia, 173
Japan, 32
jellyfish, 15–16
Jenkyn, Thomas, 54
jet contrails, 7, 81, 139
Jinan, China, 192
Job 38:16–36, 131
Johnson, Lyndon B., 221
Jurassic, 226

Keith, David, 23, 34, 218
Kelemen, Peter, 201–4, 208, 214, 215,
    216, 220, 223, 227, 229, 230
Kenya, 142, 152
Kharitonchik (rock), 242
Khosla, Vinod, 145–46, 148
King, F. H., 82
King, Martin Luther, Jr., 232
Kirkland, Isabella, 105
Klamath people of Oregon, 154
KlimaFa (Climate Trees), 18
Korea(s), 51–52
Krakatoa (Winchester), 41
krill, 27, 28, 30, 32, 35, 36, 37
Kronos, 29
Ku Sim, 52

Lackner, Claire, 217–18
Lackner, Klaus, 209–10, 211, 212, 213,
    216–19, 230
Lahore, Pakistan, 175
Lake Nyos, Cameroon, 219–20
Lamarr, Hedy, 158

Lanza, Robert, 115–16
Lao Tzu, 165
LeConte, Joseph, 54
Leicester, England, 40, 41–42
Leicester University, 51, 63
Lei Hengshun, 148–49, 150
Lem, Stanisław, 54
Leopold, Aldo, 92
Les Époques de la Nature (Leclerc), 53
Liangcheng, China, 172
Liberia, 81
Library of Congress, 153, 154–55
lidar, 95
life
    belief in all life as precious, 60
    DNA source code for, 152
    edited genomes and, 115
    hardiness of, 61
    human alteration of, 6
    oxygen and, 205–6
    power of, 96
    synthetic biology and, 107
    as Weltorganismus, 54
Li Hejun, 257
Lijiang, China, 132, 133
Linnaeus, Carl, 39, 119
lions, 121
Li Shiyang, 160–61
Little Ice Age, 44
lobster, 15
Lockheed, 243
Long Now Foundation, 108–9, 121, 127
Lovelace, Ada, 158
Lovell, James, 1
Lovelock, James, 41
Lu, Sheldon, 174
Luwesi, Cush, 228
Lyell, Charles, 54, 55, 63–64
Lyubov Orlova (ship), 15

Mad Max: Fury Road (film), 53
Majumdar, Arun, 145, 146, 147, 148, 209
Ma Jun, 189
mammoths, 44, 53, 71, 101, 104, 105,
    120, 122, 126, 127

Church's experiments, 107, 115, 152
    de-extinction of, 127, 234
    DNA of, 106, 116, 117, 125
Man and Nature (Marsh), 44, 54
Mao Zedong, 133, 150
Marianas Trench, 14
Mars, 6, 224, 238, 242–45, 252
Marsh, George Perkins, 44, 54
marsupial wolf (thylacine), 101, 128
Martin, John, 28, 35
Maryland, 93
mastodon, 101, 104, 107, 112
McGrew, Mike, 126
Melbourne, Australia, 166
memes, 52
Mesopotamia, 172
methane, 44, 84, 159, 166–67, 207
Mill, John Stuart, 48
Miller, Heinrich, 31
Minuteman missiles, 259
Mississippi River, 20, 59, 142
modernity, 263–64
Moffat, Scotland, 40
mollusks, 16
Monbiot, George, 62
Monster of God (Quammen), 121
moon, 1–2, 6, 45, 244
mountain gorilla, 126–27
Mueller-Stahl, Armin, 26
Mumbai, India, 175
Musk, Elon, 2, 18, 152, 232–39, 245–49,
    253–57, 260, 261
    joke, 233
    Mars and, 238, 247, 252
    solar power and, 246, 253–56
    SpaceShip Two, 244
    space travel and, 243–45, 246, 249
    SpaceX, 234, 235, 243, 245, 247–48
    Tesla, 232–39, 245, 246, 248, 249
Musk, Kimbal, 247
musk oxen, 101

Nagoya targets for conservation, 96
NASA, 1–2, 3, 6, 77–78, 244
    Landsat satellite imagery, 81

SpaceX contract, 248
    U.S. out of the space race, 251
National Geographic Society, 104–5
national parks, 48
National Space Society, 247
natural resources, 62
    American consumption of, 136
    consumer goods and, 138–39
    Hancock Prospecting and, 257
    leveling mountains, China, 191
    technology needs, 138, 151
    Yunnan Province, 133–34
Nature Conservancy, 31
Neanderthal genome, 102
Nero, 94
Newfoundland, 201
New Mexico, 240, 249–50
New York City, 93, 175
    average garbage daily per person, 186
    Battery Park City, 186
    Gowanus Canal, 188, 190
    sewage, 188
    solar power for, 255
    waste problem, 186–87
Nigeria, 142
nitrogen, 22, 45, 59, 83, 85
    from atmosphere, 85, 207–8
    ocean dead zones and, 59, 208
North America
    agriculture and, 48
    animal lineages of, 101
    arrival of humans, 101, 103–4
    changing biomes of, 74
    disease brought by Europeans, 72
    earliest humans, 71
    European arrival, 71–72, 93
    extinction of the mammoth, 71
    Great Eastern Forest, 123
    horses in, 101
    human reshaping of, 71
    Maryland, 71–72, 93
    reforestation, 74, 80, 93, 122
    resilience of nature and, 74
    tobacco, 72
North Dakota, 97, 99

North Sea, 16
Norway, 32
Novak, Ben, 97–100, 104–12, 122, 129
    DNA sample collection, 117, 124–25
    passenger pigeon and, 97, 99–100,
        104, 106–12, 114, 118, 124,
        128–30
nuclear power, 4, 23, 138, 146, 187
    waste, zero release and, 220
nuclear weapons, 4, 65, 238–42
    Anthropocene start date, 44, 57, 65
    "bomb curve," 240
    radiation and, 241
    Trinity test site, 238–42
    U.S. Minuteman missiles, 259
NUMMI, 234–35

Obama, Barack, 59, 183
Ocean Nourishment Corp., 18
Ocean of Life, The (Roberts), 17
oceans, 6, 11–38, 184
    acidifying of, 16–17, 20, 29
    algae blooms and, 20, 184
    changes in bluefin tuna, 15
    changing shorelines, 20
    chemistry of, 4
    CO_2 sequestration and, 208
    collapsed ecosystems, 15
    coral as indicator of health, 32
    cyanobacteria as food, 12
    dangers in, 14
    dead zones, 59, 208
    dimethyl sulfide (smell of the sea), 20
    extinctions, 14, 15
    farming of, 20–21, 36
    first global map of, 14
    food chain, changes in, 15
    as garbage dump, 5, 14, 32, 189
    human ignorance about, 14–15
    humans as evolutionary force, 32
    iron fertilization experiments and,
        13–14, 17–20, 28, 29, 30, 32–37
    killing jellyfish eaters, 15
    killing of coral, 14, 16
    krill restoration efforts, 37

oceans (*cont.*)
loss of bottom life, 16
melting ice sheets, 26, 56
missing ships, 15
Nagoya targets for conservation, 96
new ecosystems, man-made, 16
nutrients carried by the Amazon, 37
overfishing, consequences of,
15–16
percent of the globe, 8, 14
photosynthetic organism, 14
plankton in, 12–13, 17–20
plastisphere, 16
rising sea levels and, 56, 222
rogue waves, 11
shifting baselines and, 15
ship tracks, 81
stewardship of, 31
tin-based coating on boats, 16
trawling or dredging of, 16
warming, 15, 16
Oligocene epoch, 220
olivine, 202, 215, 227, 229, 230
Olsen, Paul, 209, 226, 227
Oman, 201–2, 204, 209, 214, 229
Oppenheimer, J. Robert, 239
Orbis spike, 44
Orbital Sciences, 243
Ordovician Age, 39–40, 65
Origin of Species, The (Darwin), 114
oryx, 241
oxygen
Amazon rain forest and, 93
Archaean Earth and formation of,
21
catastrophe and extinction, 22
percentage of atmosphere, 13, 205
photosynthesis and, 205
plankton and creation of, 13, 93

Pääbo, Svante, 110, 111
Pacific Ocean, 31
Page, Larry, 152, 257
paleogenomics, 102, 107, 114, 117
Palisades, New Jersey, 225–27

Palmyra atoll, 31–32
Palos Verde blue butterfly, 106
Panama, 89
Panama Canal, 28, 90
Pangaea, 27
Paraguay, 81
Paris Agreement, 220, 223, 229, 260
Park, Ah-Hyung, 208–11
passenger pigeon, 97, 99–100, 110–14,
124. *See also* Novak, Ben
date of extinction, 112
de-extinction of, 100, 106–9, 113,
114, 118–19, 126, 127, 128–29
habitat revival of, 122–23
Long Now funding, 108–9
Martha, last of, 112–13, 114
massacre of, 1847, 111
as "outbreak species," 112
Pavlov, A. P., 55
PayPal, 246, 247
Pearce, Fred, 137
peridotite, 202, 203, 204, 209, 214
Permian period, 45
pesticides, 85, 139, 168, 179
Phelan, Ryan, 108, 121
Philips, 143, 144
photosynthesis, 13, 21, 37, 47, 123, 205,
209, 220, 223
Pinatubo volcano, Philippines, 224
Pinta Island giant tortoises, 112
Piscataway tribe, 72
Pittsburgh, Pennsylvania, 179
Planet of the Apes (film), 4
plankton, 12–13, 18–20, 28, 29, 37
atmospheric composition and, 13,
93
diatoms, 12, 29, 35, 36, 37
iron fertilization experiments, 17, 18,
19–20, 29, 32–37
Planktos Inc., 18, 19
plants. *See also* trees
climate change and, 62
Columbian Exchange, 62
de-extinction of, 123
evolution of, 22–23

invasive species, 92
  managed relocation, 62
  speeding up of extinctions, 61
Plass, Gilbert, 221
plastic, 6, 14, 32, 46
plastiglomerate, 55, 57
plastisphere, 16
plate tectonics, 7
Pleistocene, 4, 26, 43, 71, 101, 264
  animals of, 71, 103, 107
Pliocene, 220
Pluto, 46
Poinar, Hendrik, 106, 112, 117, 127
Polarstern (ship), 12, 13–14, 18, 27, 29, 32–33
Poole, Robert, 244
population growth, 136–37, 208
potato, 62
Powell, William, 123
Principles of Geology (Lyell), 54, 63–64
Principles of Political Economy (Mill), 48
"providential principle," 62
Psychozoic age, 54

Quammen, David, 63, 121
Quaternary, 46, 55
Queen Mab (Shelley), 69

Ragland, The (ship), 18
rain gardens, 253
rainmaking, 23, 24
Ramshaw, Gary, 250
"rapid iteration," 260
Rare (environmental group), 160
rats, 61, 137
Reagan, Ronald, 183
redwoods, 123
Rees, Martin, 121
religion, 5, 53, 55, 139, 171, 176, 199, 249, 262
Renaissance, 92
Renninger, Neil, 145
Révoil, Benedict Henry, 111, 113
rewilding, 120–21

Rinehart, Gina, 257
Rive, Lyndon and Peter, 246, 253, 254, 256, 260–61
Rizhao, China, 165–72, 182–87, 191–96, 199
Roberts, Callum, 11, 17
Robinson, Kim Stanley, 53
robotics/robots, 86–87, 96, 242–43
Rochert, Ralf, 34
rock dove (*Columbia livia*), 118
Rodale, Robert, 89
Romania, 121, 122
Roswell, 240
Rowell, Galen, 3
Rubin, Rick, 237
Ruddiman, William, 48
Russia, 23, 242, 251

saber-toothed tiger, 71, 101, 104
Sagan, Carl, 7, 263
Saintaugustinean, 101
salmon, 30, 36
salps, 36
Sandia National Laboratories, 146
Santarosean, 101
Santayana, George, 189
satellites, 80–81
Saudi Arabia, 256
Sax, Dov, 127–28
Schavan, Annette, 17–18
Scheele, Carl Wilhelm, 205
Schneider, Stephen, 24
Schuiling, Olaf, 214, 215
scientific method, 64, 243
scientism, 187–88
Sci Foo, 151–53
sea cucumbers, 16
Sea Shepherd Conservation Society, 19
Seattle, 94
sea turtles, 15
Sen, Amartya, 82
Seneca, 154, 155
serpentine, 230
Seto, Karen, 60
sewage, 31, 84–85, 184, 188–90, 223

Shanghai, China, 178, 179, 181–82, 189–90, 200

Shapiro, Beth, 102–4, 107–9, 116, 117, 125, 128, 129

Shelley, Percy Bysshe, 69

Shenyang, China, 179–81, 191

Shigu, China, 150

Shijialing village, China, 165, 167–68

Shi Tongkang, 167

shrimp farming, 36

Siberian Traps, 227

Sierra Leone, 81

Singapore, 173

Skycatch, 80

Sleipner, 213, 227

Smetacek, Fritz, 11

Smetacek, Karen, 24–25

Smetacek, Victor, 11–14, 24–25, 28, 35, 38, 203
    iron fertilization experiments, 17, 18, 19–20, 29, 32–37

Smithsonian Environmental Research Center (SERC), 73, 88

Smithsonian Institution, 54, 113, 183

smog, 23, 81, 142, 171, 176, 178, 189, 192, 193, 194, 196, 198, 208

Snowpiercer (film), 53, 224

Snyder, Gary, 86

Social Conquest of the Earth, The (Wilson), 97

soil erosion, 59, 142

SolarCity, 238, 245, 248, 253, 261

solar power, 142, 144, 148, 154, 183, 199, 246, 254
    Chinese photovoltaic production, 183
    cost of, 184, 253–56
    electricity production, 183
    for hot water, 157, 183
    making solar panels, 212
    Musk and, 246, 253–54
    Saudi Arabia and, 256
    on the White House, 183

Solar Science and Technology Museum, Dezhou, 183

Song of the Dodo, The (Quammen), 63

soot, 20, 45, 65, 81, 134, 138, 143, 170, 178, 187, 189, 195, 220, 254

South Africa, 142
    Aldinville School, 157, 158
    China as future model, 157–58
    coal in, 142, 144
    cooking smoke, 143
    electricity, 140–41, 144, 158
    iLembe District, 140, 144, 156–57
    modern energy and health, 156–58
    wind turbines, 144–45

Southern Ocean, 11–14, 19, 28–37

South Korea, 210

space, 3, 6–7, 232–66
    Apollo 8 mission, 1–2, 6, 263
    Branson and, 243, 249–51
    Challenger space shuttle tragedy, 244
    CO_2 capture and storage and, 244
    cosmoscene and, 263
    escaping Earth and, 252–53
    human adaptation and, 261
    Lunar Orbiter 1, 244
    Musk and, 242–49
    pictures from, 1–2, 3, 263
    race for Mars, 242–43
    satellites in orbit, 80–81
    travel, 152–53, 238

Spaceport America, 249–50

SpaceX, 234, 235, 243, 245, 247–48, 250, 251, 255, 260

Spanish flu, 140

Speaking for Nature (National Audubon Society), 99

squid, 15

Star Wars (film), 251

Steller's sea cow, 100

Stephans, Joseph, 112

Stephans, Salvator, 112

Stereocene, 264

Steyer, Tom, 257

Stiller, Matthias, 111, 116–17, 120

Stoermer, Eugene, 42, 52, 53

Stoppani, Antonio, 54

stratigraphers, 40

strontium 90, 45

sulfur, 21, 22
Sumeria, 52
Sun Microsystems, 145
Suomi satellite, 263
sustainability, 66, 89
    of energy, 141, 255
    trees and, 131–35
synthetic biology, 107, 121, 122
Syvitski, James, 59

Tambora volcano, Indonesia, 224
Tao Te Ching (Lao Tzu), 165
Tasmanian tiger, 100, 112
Tata, Ratan, 257
technology, 105
    to capture CO_2, 211–20, 228–29
    existential risks of, 66
    Laudato Si' encyclical, 258
    natural resources used for, 138
    negatives of, 7
    nuclear weapons, 242
    space and space travel, 244
techno-optimism, 85, 108, 153
TEDx event, 105, 108, 109, 122, 126
Tertullian, 63
Tesla, Nikola, 160
Tesla Motors, 232–39, 245, 246, 248,
    249
Thomas, Jim, 17
3-D printers, 153
Tianjin, China, 181, 182, 184, 198
Tibet, 133
tobacco, 72
Tokyo, Japan, 173
trains, high-speed, 191–92
Trans Am (band), 53
trees
    agriculture and loss of, 80, 81
    Amazon rain forest, 15, 23
    American chestnut, 123
    carbon dioxide stored in, 95
    carbon neutrality and, 170
    of China, 168, 178, 179, 187
    Chinese proverb, 161
    number of, worldwide, 80

planting, to reduce CO_2, 18
    sustainable practices and, 131–35
    tulip poplar, 95
Triassic, 226
Trinitite, 240, 241–42
tuna, 15
Turner, Ted, 249
Twain, Mark, 24

Union College, Maine, 183
United Arab Emirates, Mars and, 243
United Nations
    Africa's energy poverty and, 140, 141
    Climate Change Conference, 222–23
    Environment Programme, 81, 200
    treaty on the Prevention of Marine
        Pollution, 36
University of California–Santa Cruz,
    102, 104, 107, 109, 110, 129
University of Iowa, Upper Holocene
    Boundary Commission, 43
University of Maryland, Baltimore
    County (UMBC), 70, 75, 77
    Herbert's Run, 75
    inventorying trees, 75–76
    the Knoll, 70–72, 74–76, 79, 94–95
Uruguay, 81
U.S. Department of Energy, 227, 239
Utsira Formation, 220

VanderMeer, Jeff, 53
van Leeuwenhoek, Anton, 137
Venus, 214, 224
Vernadsky, Vladimir, 44
Virgin Earth Challenge, 215
Virgin Galactic, 243, 251
viruses, 116–17, 126
volcanic eruptions, 30, 45, 49, 205, 206,
    224, 227
Vonnegut, Bernard, 23
Vonnegut, Kurt, 23

Wake Forest, 81–82
Walkowicz, Lucianne, 251–52
War Games (film), 259

wars, innovation and, 51–52
water. *See also* oceans
 for agriculture, 139
 chlorophyll and, 21
 collection, Zululand, 157
 dams, 3, 5–6, 94, 148–51
 fertilizer and "dead zones," 20, 30
 rivers reshaped by people, 59
 Rizhao's plan, 184
 wasted, 138
water pollution, 132, 178, 179, 184,
 188–90
wealth inequality, 177
weather, 15, 20, 23, 24. *See also* climate
 change
Weatherbird II (ship), 19
Wegener, Alfred, 27
Wen Jiabao, 188
West, Geoffrey, 151
western meadowlark, 97
Wetlands International, 121
whales, 15, 16, 19, 27, 30, 35, 117–18
 decline of cetaceans, 29–30
 restoration efforts, 37
white rhino, 122
White Sands Missile Range, 240, 250
wildfires, 59
wildlands, 8, 62, 70, 86
Wildlife Conservation Society, 123–24
Williams, Mark, 63, 64–65
Williston, North Dakota, 97, 98–99
Wilson, E. O., 96, 97, 130
Winchester, Simon, 41

wind power, 25–26, 34, 142, 144, 148,
 176–77, 180, 187
wolves, 113, 120
woodland caribou, 106
World's Fair of 1964, 107
World War II, 49–50, 52
Wright, Allen, 218, 219

Xerces blue butterfly, 105
Xi Jinping, 188, 191, 193, 197
Xinjiang, China, 177

Yan Hui, 185
Yeha, Ethiopia, 173
Yellow River, 59
Yellow Sea, 165, 181, 184
Yellowstone Park, 73
Young, Neil, 18

Zalasiewicz, Anna, 51
Zalasiewicz, Feliks, 51
Zalasiewicz, Irena Beresniewicz, 49–50
Zalasiewicz, Jan, 39–41, 49–51, 53, 55,
 56, 58, 60, 64, 92
 Anthropocene marker, 40–44, 46
 Anthropocene start date, 57, 65
 who he is, 42
Zhang Guobao, 196
Zhenjiang, China, 199
zooplankton, 12, 36
 copepods, 20, 36
Zulus, 140
Zuma, Jacob, 140, 141, 144

About the Author

David Biello is an award-winning journalist who has been reporting on the environment and energy since 1999. He is currently the science curator for TED as well as a contributing editor at *Scientific American*, where he has been writing since 2005. He also has written for publications ranging from *Aeon* and *Foreign Policy* to the *Los Angeles Review of Books* and *The New Republic*. He hosts the ongoing duPont-Columbia award-winning documentary *Beyond the Light Switch* as well as *The Ethanol Effect* for PBS. He lives in Brooklyn with his wife and two children. *The Unnatural World* is his first book.

per year, to take just one example. The technofossils of former mines will be ubiquitous and, perhaps, some bright future geologist will make the connection: coal removed here and then dumped back in as ash, copper removed there and accumulated here.

In other words, those future geologists will have to puzzle out well bores filled in by collapsed rock and plastiglomerates, and then all the things people did to the surface of the land, from changing river sediment flows with all those dams to shifting cement and stone to raise modern cities. "The thing with the Anthropocene is that it has become even more of a kind of intricate filigree, if you like," Zalasiewicz goes on. "So you have the urban strata. You have other buildings and the foundations and all the rubble that moves around. That is put on geological maps, but it is in pockets and patches and pods, and of course you have this in all the pipe work in the sewers and transportation systems, and so it's an extension. It's a terrifically complex geometry of physical objects and deposits, and that goes in the ground as well."

This filigree is unique in geologic history, a special mark in the more than 4 billion years of Earth's life preserved in rock. "There's nothing which forms these kinds of particular patterns which has emerged before this in the last 4.5 billion years. We're creating new physical patterns."

The new epoch may prove as obvious as the Cambrian boundary, when burrowing started in the seafloor. Deep-sea trawling may leave a similar record for the far future, and, certainly, the land has been marked. "Since ancient cities show up well in archaeological excavation, this spate of urbanization will be evident stratigraphically in the distant future," Zalasiewicz and his coauthors wrote in a paper formally arguing for a nuclear start to this new epoch.

Perhaps it is waste that will last, nuclear or otherwise. In fact, some might argue that the present era begins with garbage, when a dense enough agglomeration of humans came together in one place that their waste—both feces and offal—began to pose a risk. Garbage is the necessary corollary of modern consumption: If you can't throw away, there is no room for new stuff, which later will become more garbage . . . Our trash will tell something of the speed of modern life with its excess of coffee cups, plastic bags, and packaging.

The first humans produced no waste other than the bodily excretions common to all other animals. These hunter-gatherers eventually mastered fire, making ash the first waste—and the one by which ancient exploits are dated through modern science (though that dating will soon be obscured by all the fossil carbon dumped in the atmosphere). The first villages and cities mark the beginning of significant waste, bodily excretions and ash now accompanied by broken bits of technology like pottery shards and discarded food. This waste can prove a rich resource, and the ragpicker probably arose as an old, old profession, one to rival prostitution in antiquity.

Millennia later, modern civilization replaces the ragpicker with the professional sanitation worker, whose efficiency requires a dump rather than the organic recycling practiced by the poor. Industrial recycling has nothing on the reuse required by poverty. The future may require a return to that logic, a circular economy with little or no waste. An enduring human civilization may require less of a rich record to be left for those future archaeologists and geologists.

"It is hard, as humans, to get a proper perspective on the human race," Zalasiewicz writes in *The Earth After Us*. This new geologic epoch threatens to put people back in the center of the action, the collective protagonist of a planetary drama. That reverses hundreds of years of science steadily whittling away at the idea of humans as central, or even special. As an idea, the Anthropocene represents the makers of geology inscribing themselves into their own rock record. People alive today may have lived in two different geologic epochs. It is an epoch-alypse, just not the one we were promised.

A voluminous correspondence builds the case for a new geologic epoch among the members of the Anthropocene Working Group, even as some incline toward the apostasy of no formal definition. And other groups find merit in the use of the idea as a lens with which to view other problems anew, like the International Geosphere-Biosphere Programme, the kind of unwieldy name preferred by overliteral scientists, a group that convenes meetings in academic outposts around the world to discuss how this new epoch might enable better urbanization, better use of nitrogen, or control of fire, as well as how to install some kind of governor, or at least ameliorate the worst effects of the speed of the so-called Great Acceleration that has wrought change on a global scale.